T0140394

Advances in Information Security

Volume 85

The purpose of the *Advances in Information Security* book series is to establish the state of the art and set the course for future research in information security. The scope of this series includes not only all aspects of computer, network security, and cryptography, but related areas, such as fault tolerance and software assurance. The series serves as a central source of reference for information security research and developments. The series aims to publish thorough and cohesive overviews on specific topics in Information Security, as well as works that are larger in scope than survey articles and that will contain more detailed background information. The series also provides a single point of coverage of advanced and timely topics and a forum for topics that may not have reached a level of maturity to warrant a comprehensive textbook.

More information about this series at http://www.springer.com/series/5576

Matthew Ryan

Ransomware Revolution: The Rise of a Prodigious Cyber Threat

Matthew Ryan
Cyber Risk and Technology Researcher
Maroubra, NSW, Australia

ISSN 1568-2633 ISSN 2512-2193 (electronic)
Advances in Information Security
ISBN 978-3-030-66585-2 ISBN 978-3-030-66583-8 (eBook)
https://doi.org/10.1007/978-3-030-66583-8

This Springer imprint is published by the registered company Springer Nature Switzerland AG
The registered company address is: Gewerbestrasse 11, 6330 Cham, Switzerland

Foreword

In 2017, the world was rocked by ransomware. First, in May, the WannaCry attack hit hundreds of thousands of computers around the world, severely impacting the United Kingdom's National Health Service, among many other organizations. Then, the following month, the NotPetya ransomware decimated computer systems at major multinational companies including Mondelez, Merck, and Maersk, among many others. People who had never even heard of ransomware before and organizations that had never included online extortion as part of their cyber risk assessments suddenly realized that this was a major threat being distributed by serious adversaries with devastating consequences. That realization has only grown since 2017 as an increasing number of organizations in both the private and public sectors have fallen victim to ransomware, forcing the temporary shutdown of municipal services, hospitals, schools, and many other institutions.

In this timely and compelling book, Matthew Ryan takes a historical look at ransomware, going back all the way to the 1980s, and traces the development of cryptographic technologies and cryptocurrencies alongside the rise of ransomware as a threat tactic. In doing so, he illuminates the different technological and societal factors that have contributed to the success of ransomware as a business model for cybercriminals. Interwoven with this historical analysis, Ryan provides a look at the current costs of ransomware, the toll it has taken on our infrastructure and operations, and the mechanisms available to us for trying to mitigate its impacts.

Combating ransomware has long been a complicated and controversial area for cybersecurity scholars and practitioners. Recommendations such as urging organizations and individuals to create regular back-ups of their computer systems to allow for full reboots when necessary may be helpful in some cases, but are impractical and unwieldy in others where it may not be possible to achieve rapid recovery even if back-ups are available. For years, law enforcement and cybersecurity professionals have struggled with what guidance to give victims of ransomware. Famously, in October 2015, FBI special agent Joseph Bonavolonta told an audience at a cybersecurity conference in Boston that the FBI often advises victims "just to pay the ransom" because "the ransomware is that good." The FBI later back pedalled from that position, insisting that their official guidance to ransomware victims is not to give into the attackers' demands, but even so, Bonavolonta's point remains an important one: that the quickest—and sometimes even the cheapest—way to restore affected systems is to pay the ransom. As insurers increasingly offer coverage for such extortion payments, firms may be even more inclined to make them, further fuelling the criminal industry that develops and distributes such attacks, and contributing to the continued profitability of ransomware that encourages even more threat actors to undertake similar schemes.

Ryan provides an unflinching look at the motives for these attacks and the psychological dimensions of their perpetrators, applying rational choice theory, expected utility theory, and prospect theory, to disentangle the complicated range of factors that underlie online extortion. His socio-technical exploration or ransomware case studies from the past decade, ranging from CryptoLocker to WannaCry, NotPetya, and Locky, reveal new dimensions of these attacks, their impacts, and the attribution process that took place in the aftermath of each one. Ryan makes clear how much the effectiveness our efforts to prevent ransomware and mitigate its effects depends on the particular psychology and motivations of the people behind those attacks and the larger goals driving them.

Ransomware Revolution provides an important addition to the growing body of work on the rise of ransomware and online extortion, the technological factors that shaped this type of cybercrime, the people who perpetrate it, and the larger, global efforts that have largely failed to prevent it. Ryan's thoughtful and thought-provoking analysis and historical perspective contribute to a comprehensive and insightful interdisciplinary exploration of different ransomware incidents that will be accessible to readers at every level of technical expertise and experience with cybersecurity. By bridging technical, economic, psychological, and political elements of ransomware, this book brings together many of the different forces that have driven the rise of ransomware in the past decade and that we will continue to grapple with for many years to come.

Cambridge, MA, United States Josephine Wolff

Acknowledgements

Undertaking research to write a book is a challenging and time-consuming process that relentlessly pushes the author to explore new frontiers. The research development process was both demanding and rewarding, and one that could not have been completed in isolation, therefore I wish to take the opportunity to acknowledge and thank the following people who have supported me throughout this journey. Firstly, Professor Greg Austin, thank you for agreeing to take me on as a student, for your timeless devotion to the project, and for rationalising all my crazy hypothesises along the way. I could not have asked for a better scholarly role model and you have been a tremendous mentor and influence over many years. I am truly grateful for your involvement in not just this research project, but also for the ongoing development of my professional career.

Completing a research project of this magnitude is certainly not a solo effort and required the advice and expertise from many world-class scholars. Dr Josephine Wolff, your research inspired me throughout my research journey, and your review generated ideas that were invaluable to the advancement of the research. Dr Herb Lin and Dr Yenni Tim, I am extremely grateful for your generosity in undertaking peer reviews and for your critical guidance throughout the research process. I would also like to thank Dr Bob Ormston, GPCAPT Andrew Gilbert, and Dr Frank den Hartog for all your tutelage in reaching this point. Additionally, I would like to offer a special thanks to Matt Combe, Bou Waterhouse, and David Kennedy for your advice and support throughout an extensive research and development process.

Finally, and most importantly, I would like to thank my beautiful wife and partner Lisa Ryan. Thank you for encouraging me to chase my dreams, your patience, and for all the amazing meals along the journey. It would not have been possible to complete this project without your unwavering love and support. Quite simply, I could not have written this book without you by my side.

Contents

List of Acronyms

ABS	Australian Bureau of Statistics
ACIC	Australian Intelligence Commission
AES	Advanced Encryption Standard
AFP	Australian Federal Police
AI	Artificial Intelligence
AML	Anti-money laundering
API	Application Process Interface
ASD	Australian Signals Directorate
BCP	Business continuity plan
BTC	Bitcoin
CaaS	Crime-as-a-Service
DARPA	Defense Advanced Research Projects Agency
DDoS	Distributed Denial-of-Service
DES	Data Encryption Standard
DGA	Domain Generation Algorithms
DHS	Department of Homeland Security
ETC	Ethereum
FBI	Federal Bureau of Investigation
GCHQ	Government Communications Headquarters
ICANN	Internet Corporation for Assigned Names and Numbers
ICO	Initial Coin Offering
IDS	Intrusion Detection Systems
IIoT	Industrial Internet of Things
IoT	Internet of Things
IPS	Intrusion Protection Systems
ISACA	Information Systems Audit and Control Association
ISO	International Organization for Standardization
IT	Information Technology
JSCoRE	Journal of Sensitive Cyber Research and Engineering
KYC	Know Your Customer
LTC	Litecoin

ML	Machine Learning
NATO	North Atlantic Treaty Organization
NHS	National Health Service
NSA	National Security Agency
OT	Operational Technology
PII	Personal identifiable information
RaaS	Ransomware-as-a-Service
RingCT	Ring Confidential Transactions
RSA	Rivest, Shamir, and Adelman
SOF	Special Operations Forces
TOC	Transnational Organised Crime
Tor	The onion router
VPN	Virtual Private Network
WHO	World Health Organization
XMR	Monero
XRP	Ripple

List of Figures

List of Tables

Chapter 1
Introduction

The first known incidence of ransomware is the AIDS Trojan developed and distributed by Dr. Joseph Popp in 1989 (Mungo and Clough 1992). Popp was a medical researcher who distributed 20,000 floppy disks to fellow researchers who attended the 1989 World Health Organization's (WHO) conference on AIDS. Popp claimed that his program could analyse an individual's risk of acquiring AIDS; however, the disk he distributed to his colleagues also contained a malware program (Trojan) that activated after the victim's computer was rebooted 90 times.[1] Once the malware was initiated, all the user's files and directories were encrypted in the computer system's root directory. Once the encryption process was completed, the malware displayed a message asking the user to pay $378 (USD) for renewing a license which could recover the lost files and directories. The request for payment asked the user to mail the ransom payment in the form of a cashier's cheque to a post office box in Panama. The trouble and time delay in paying the ransom and receiving the decryption key ultimately limited the profitability of the attack (Solomon et al. 2000). It is unclear how many people fell victim to the attack; however, it didn't take long for the Federal Bureau of Investigation (FBI) to attribute the attack to Popp, and he was arrested in February 1990 and extradited from London to the United States to face trial.[2]

Whilst this is the first known instance of a ransomware attack, it required manual delivery of the malicious software and physical payment of the ransom. In the following decade, the technique was rarely employed by cyberattackers. However, this initial nonchalance would begin to change with the rapid spread of the Internet,

[1] Note: The AIDS Trojan was named by its developer Dr. Joseph Popp, who attempted to defend his unscrupulous actions by arguing the ransom payments were to be used to undertake further medical research into AIDS. See also Anonymous (1998).

[2] Note: At the hearing, Popp was ruled mentally unfit to stand trial and remained free until his death in 2007. See Simone (2015).

M. Ryan, *Ransomware Revolution: The Rise of a Prodigious Cyber Threat*, Advances in Information Security 85, https://doi.org/10.1007/978-3-030-66583-8_1

Internet-enabled devices, new technology platforms and unrestricted access to advanced encryption methods. The advent of the Internet meant there was no longer a need to physically hand out infected disks anymore. Ransomware attacks could now be developed, purchased and launched remotely. It also enabled ransom demands to be requested in anonymous cryptocurrencies and online payment platforms.

Despite a burgeoning list of victims that were well resourced, corporations continue to forge ahead with a path that is congenial with ransomware. For corporations, the threat posed by ransomware has emerged from a collective of cyberthreats, rising to become the most serious threat that organisations confront. Ransomware attacks grab our attention, but the reality is organisations learn little from them, instead of focusing their finite resources towards broader cyberattacks and deploying emerging technologies. To further understand the threat ransomware poses, there is an underlying requirement to better understand what has happened and how we got to this point.

Throughout the period of globalisation, governments, corporations and the public have rapidly adopted new computer-based systems, software platforms and Internet-enabled devices. The global market for consumer technologies and personal electronic devices continues to grow exponentially through the continuous product development cycles for new Internet-enabled devices, mobile phones, Internet of Things (IoT) devices and Internet-enabled Operational Technologies (OT). Whilst consumers have been quick to adopt these emerging technologies, they have generally been slow to recognise the security-related threats associated with these emerging technologies and platforms. An underlying trait of the cybersecurity discourse is that applying the necessary security controls is inherently considered to be a reactive, not proactive, process. Natively, organisations have large attack surfaces with complicated internal structures and processes, whereas cybercriminals are astute and positioned to rapidly exploit security control deficiencies and vulnerabilities.

As a result, cybersecurity has become a national priority for governments, corporations and ordinary citizens. Globally, governments, law enforcement agencies, corporations, academic institutions and the media continue to highlight the significant threats posed by cyberattacks. Despite these ominous warnings, cyberattacks not only continue to occur, but they continue to succeed at unprecedented rates. The continued success of organised cybercrime syndicates suggests that cyberattackers may be so agile and dynamic that they can't decisively be deterred, prevented or countered by conventional government defences and law enforcement agencies. In contrast to conventional government defences and law enforcement agencies, cybercriminals' agility enables rapid adoption and deployment of advanced new technologies and techniques.

This research characterises the speed of technology adoption as a fundamental factor in the continued success of financially motivated cybercrime. For businesses and criminals alike, new technologies create new opportunities. Whilst it can be difficult to measure the true speed of technology adoption for governments and law enforcement agencies globally, it is possible to broadly establish consumer adoption speeds through new device sales in developed and emerging markets. This approach is applied by DeGusta (2012) whose research indicates that mobile computing devices are being adopted at unprecedented speeds (DeGusta 2012). The research also identifies the cost of entry as a driving force behind the current and future speed of technology adoption. The EastWest Institute has also revealed the relative speed with which different groups can adopt or respond to the new technologies, and the concept they developed is shown in Fig. 1.1, illustrating in very broad terms the relative speed of technology adoption by group. Intrinsically, since law enforcement agencies must be funded by government, they will always move more slowly than government leaders. Government leaders will always move more slowly than corporations since the former have better access to expertise and more exposure to criminal risk. Criminals always move technologically ahead of corporations because they hunt out the vulnerabilities of corporations and governments.

This disparity in technology adoption identified by the EastWest Institute is supported by cybersecurity research undertaken by the RAND Corporation (EastWest Institute 2011). When researchers examined the correlation of cyberattacks and the speed of technological adoption, they found that the ability to attack will likely outpace the ability to defend. Attackers can be hedgehogs (they only need to know one attack method, but do it well), whilst defenders must be foxes (they need to know everything, not just technical knowledge but knowledge of networking, software, law enforcement, psychology, etc.) (Ablon et al. 2014a, p. 31). This underlying concept of rapid adoption is a primary reason that ransomware has the potential to remain a long-term cyberthreat to governments, corporations and individual user systems.

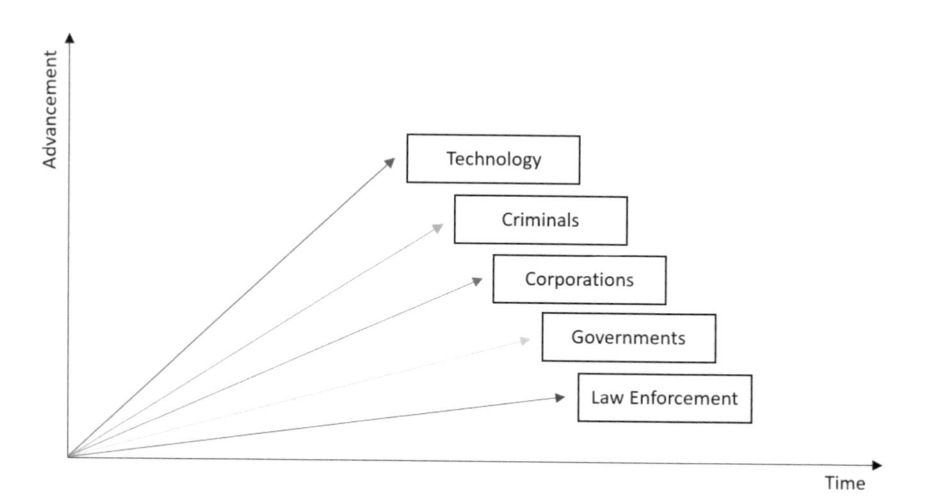

Fig. 1.1 Speed of technology adoption

1.1 The Problem

In spite of a high degree of public awareness, and in spite of an increasing number of advanced cybersecurity countermeasures being released to the market, ransomware poses a greater existential threat to corporations than other forms of cyberattack do. Coburn et al. (2018) argue that ransomware has become one of the most pressing concerns for cybersecurity specialists, and attempts to extort major companies using cyberattacks have grown in frequency, scope and ambition (Coburn et al. 2018, p. 16). Moreover, even if a ransomware attack does not cripple a corporation's Information Technology (IT) operations, it can still pose a serious threat to the data contained within them. This threat extends beyond corporations and applies equally to IT and OT infrastructure and data owned and operated by governments and individual IT users.

Research by scholars and other IT professionals undertaken throughout the past decade have provided some answers on how to prevent, to detect and to respond to certain forms of ransomware attacks. But the literature and practice to date have been largely concerned with how to defeat known ransomware attack methodologies. There is still no practical and cost-effective method to prevent successful ransomware attacks, particularly one using novel approaches (either technological or psychological). This research hypothesises that should the phenomenon of ransomware be proven to be a dynamic threat, then the academic literature and industry best practices towards preventing and defending ransomware attacks must be dynamic and adaptive to respond to the phenomenon.

For Transnational Organised Crime (TOC) syndicates, cyberattacks have traditionally been considered to be of low risk and highly profitable ventures due to the inherent attribution problems associated with cyberattacks. From the Internet's outset, cybercriminals typically focused on cyberattacks that were aimed at stealing financial information, disrupting or denying a user's access to a system and an array of cyberattacks designed to embarrass their intended targets. Whilst these types of cyberattacks continue to occur frequently, in the early twenty-first century, there was dramatic shift in modus operandi from stealing a user's sensitive information to deploying ransomware to deny users access to it. The timeline in Fig. 1.2 (fold-out) illustrates the emergence of the first major ransomware attacks in 2013, which by 2016 had spawned into a global epidemic for government and enterprise computer systems (Albrecht 2017).

This shift triggered a phenomenal rise in the scale, complexity and volume of global ransomware attacks. It is this dramatic shift in modus operandi that requires further examination to determine what caused the fundamental shift. Was this shift simply the next evolutionary step forward in cyberattacks? Or was it triggered inadvertently by the rapid diffusion and adoption of emerging technologies? To begin answering these questions requires examination of multiple disciplines in order to further understand the cyber environment in the lead up to the ransomware phenomenon. Empirically, it can be argued that most financially motivated cyberattackers have sought to undertake discreet criminal operations that avoid drawing the

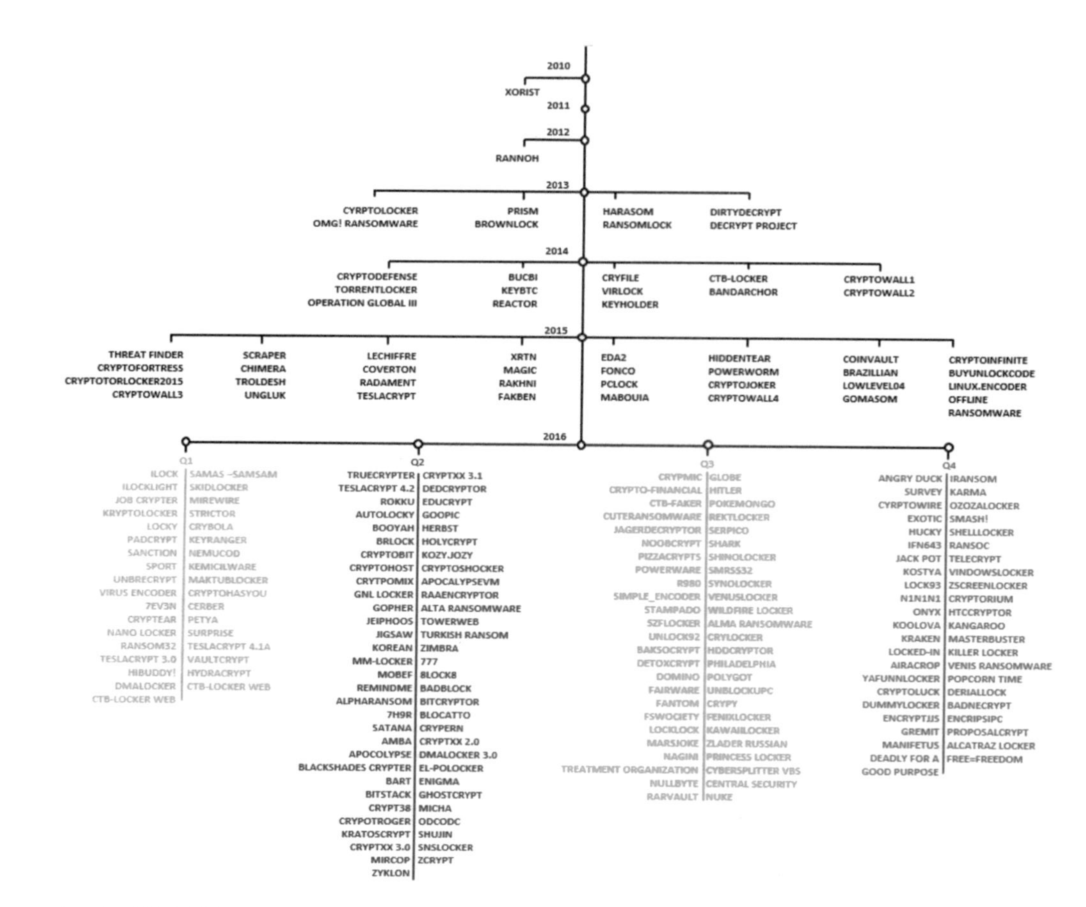

Fig. 1.2 Ransomware timeline 2010–2016

attention of powerful governments and corporations. The rise of ransomware challenges this paradigm, with ransomware attackers not only willing but, in multiple cases, deliberately targeting powerful and well-resourced opponents. The ransomware scene also changed fundamentally around 2013 as it became clear that governments themselves were prepared to create and/or use ransomware for their own purposes, leading, for example, to the globally damaging attacks in 2016 that have been formally attributed to Russia and North Korea (Bossert 2017; Taylor 2018).

One of the largest impediments to producing research on ransomware attacks is the lack of quality source data available. Holistically, the cybersecurity field is a dynamic and emerging academic discourse with a very limited number of researchers producing high-quality academic research. As a result, there is an extremely limited catalogue of research on ransomware attacks and broader cyberattacks. Austin and Slay (2016) argue that "there is little evidence that there is a generally held academic model, or body of knowledge, that applies to the cybersecurity profession and beyond that to cyber defence or cyber war" (Austin and Slay 2016, p. 16). This knowledge deficiency is supported by Knott (2014) who reasons "even for those in the cyber security community who support the need for a science of cyber, the exact nature of the new science, its scope and boundaries remain rather unclear" (Knott 2014).

The absence of high-quality data and extensive data sources is a significant impediment to researchers' undertaking academic research in the field. Moore et al. (2019) found that in most cases, data is collected in "an ad hoc, one-off fashion, requiring special arrangements with source companies. The resulting datasets are not further shared. This makes reproduction or replication of results somewhere between difficult and impossible, hindering scientific advances" (Moore et al. 2019, p. 2). This is also problematic from a researcher's perspective as academic and industry experts are often inexperienced, not independent and constricted by contracts and hail from a diverse array of academic fields. As will be discussed later, this shortcoming is prevalent throughout the literature on ransomware, with many distinguished scholars in alternative fields contributing solitary papers to the field of cybersecurity. However, it needs to be noted at the outset that most research in cybersecurity that goes beyond analysis of code is hampered by severe data limitations. The US Defense Advisory panel argues that "cybersecurity research is handicapped by the lack of scientific understanding of the cyber phenomena, particularly the fundamental laws, theories, and theoretically grounded and empirically validated models" (JASON 2010).

It should also be acknowledged that within academia, there are small groups of researchers that are producing high-quality cutting-edge research. However, many of these researchers are employed or funded by defence and intelligence agencies; thus, their research is not open source.[3] One example of this is the Journal of

[3] Note: Within the United States, the Westpoint Military Academy and the Naval Postgraduate School are examples of institutions that produce advanced cyber security research. However, by default this advanced research would be classified information, and therefore not shared with the broader cyber security research community or public.

Sensitive Cyber Research and Engineering (JSCoRE), which publishes and disseminates advanced research to a small group of officials and other security-cleared researchers (Aftergood 2019). This is not a new occurrence; organisation's such as the Defense Advanced Research Projects Agency (DARPA) were producing classified literature since the 1970s. However, there are obvious limitations with undertaking classified research; researchers cannot take full advantage of the standard practice of peer review and publication process to assure the quality of their work and to disseminate their findings.

Alternatively, outside academia, many cybersecurity researchers are employed in the professional services industries, which prohibit them from discussing intricate details of major ransomware attacks due to confidentiality agreements with their employers and clients. As a result, the research produced is retained internally by the professional services and client organisation.[4] This in turn creates silos of vast information within corporate repositories but restricts broader collaborative efforts by researchers. These data shortfalls are also compounded by the rapid and continuous evolution of ransomware attacks.

This research has two main objectives. First, it seeks to provide an original analysis of several mutually reinforcing causes for the persistent successes of ransomware attacks in spite of attention directed toward it. These include the parallel emergence of technologies such as open-source computer-enabled cryptography (encryption), virtual black markets and anonymous Internet-based payment and communications platforms. Together, these technologies have created an operating environment that has become increasingly conducive for cybercriminals to develop and undertake ransomware attacks.[5]

Second, the research offers an original assessment of these circumstances relevant to professional practice. The persistent success of ransomware attacks indicates that many corporations and governments lack the required expertise, training and effective expenditure of resources to prepare for and to respond to ransomware attacks against their organisations. The convergence of technologies has produced a cyber challenge that appears beyond the capacity of conventional risk managers and risk management frameworks, and ultimately law enforcement. The subsequent analysis of existing academic research and industry sentiment indicates these deficiencies will most likely continue to occur throughout the next decade.

The research concludes with a practical proposal for elevating ransomware defence and mitigation to the highest level of priority for corporations with a low-risk appetite for IT downtime or loss of access to data. Irrespective of whether the data is considered to be the crown jewels or whether it appears to be of little or no

---

[4] Note: Even other employees within these organisations may not be aware of the existence of internal research on ransomware attacks. Due to the potential impact to the organisation's reputation and stock prices, research data associated with previous ransomware attacks will most likely be limited to a need-to-know basis.

[5] Note: From an attacker's perspective, the gains from the emergence of these technologies are not mutually exclusive to only ransomware as these technologies are also enablers for many other forms of cyberattack.

value, ransomware can rapidly cripple and induce chaos across large enterprises. Ransomware is one of more than 30 known types of cyberattack that executives and senior risk managers need to prepare for, and recent events are quickly highlighting that it can be the most challenging form of cyberattack to prevent and recover from.

1.2 Academic Foundations

The concept of cryptolockers being used for exploitation purposes is a relatively new occurrence that can be traced back to Columbia University's Adam Young and Moti Yung's (1996) research paper *Cryptovirology* (Young and Yung 1996). The research paper presented a "set of attacks that involve the unique use of strong (public key and symmetric) cryptographic techniques in conjunction with computer virus and trojan horse technology" (Young and Yung 1996). The research was successfully able to hypothesise and later demonstrate two key elements of ransomware. The first was the researchers were able to "demonstrate how cryptography (namely, difference in computational capability) can allow an adversarial virus writer to gain explicit access control over the data that his or her virus has access to… (assuming the infected machines have only polynomial-time computational power)" (Young and Yung 1996). According to Terr (2019), an "algorithm is said to be solvable in polynomial time if the number of steps required to complete the algorithm for a given input is O (n^k) for some non-negative integer k, where n is the complexity of the input" (Terr 2019) This element of the hypothesis is explored in further detail in the sections on applied cryptography.

The hypothesis' second notion was about how "a computer virus (Trojan horse) can use a public key generated by the author to encrypt data that resides on the host system in such a way that the data can only be recovered by the author of the virus (assuming no fresh backup exists)" (Young and Yung 1996, p. 133). They argue that both of these functions are essential to an attacker executing a successful ransomware attack. Whilst the researchers offered multiple ways to potentially mitigate or recover from these types of cyberattacks, the next two decades would provide a vastly different online and digital environment.[6]

Despite Young and Yung's research results being released in 1996, ransomware did not come into the mainstream limelight until the mid-2000s. Nazarov and Emelyanova (2006) published one of the first academic research studies into ransomware (Nazarov and Emelyanova 2006). This research report provided a detailed analysis of the Trojan PGPCoder/GPCode, which was one of the first known implementations of the ransomware model proposed by Young and Yung. The research results indicated that the virus examined successfully displayed the ability to encrypt

[6] Note: It can be argued that public key encryption is not a necessary technology for ransomware attacks because there are conceiveable ways of using shared key encryption to implement a ransomware attack, even if it would be more difficult and cumbersome than with public key encryption.

more than 80 different types of files on the victims' disk drive. Another early academic research paper on ransomware was undertaken by Gazet (2010), who argues that:

> To be a successful mass extortion scheme, the ransomware must contain three properties. The first property is the malicious binary code is compromised and must not contain any secret information (e.g. deciphering keys). Secondly, the author should be the only one able to reverse the infection. And third, decrypting one device cannot provide any useful information for other infected users or devices (Gazet 2010, p. 90).

This is expanded on by Orman (2016) who argues that to successfully conduct a ransomware on a large scale, there are normally three prerequisites that need to be met, and the existing state of technology enables the attacker to meet these. The components outlined by Orman specify that:

> 1) the payload containing the hostile binary code must not possess the deciphering keys. This is where Command and Control comes into play. The idea is to not make it easily retrievable. In hindsight, white box cryptography can be applied to ransomware, 2) the cyber-criminal is the only one that should be able to decrypt the infected device and know the location of the decryption components present in the crypto files of the system, and c) the modern ransomware now mimics the properties of a worm and can self-propagate to multiple devices on the network or Work group and thus it is essential that one device cannot provide any decrypting information for other devices that get infected, which basically means that the key must not be shared among them (Orman 2016).

Despite undertaking one of the most significant early academic works into ransomware attacks, Gazet's research ultimately concluded that the "ransomware phenomenon is a reality that has to be monitored but, in some ways, it is not a mature and complex enough activity that deserves such communication around it. Ransomwares as a mass extortion means is certainly doomed to failure" (Gazet 2010). Even as late as early 2017, security and risk experts such as Brian Sims were adamant that end users should be cautious not to be caught up in the hyperbole of ransomware attacks because they still consist of very well-known components, and the responses to these kinds of threats are already well understood (Sims 2017).

The rapid evolution of ransomware attacks and associated cybersecurity threats has left governments and corporations clutching for viable and effective solutions. This quandary has drawn researchers and professionals from multiple disciplines to the cybersecurity and risk management fields. Neil Campbell, Telstra's Director of Security, explains that cybersecurity is no longer just a technology-centric issue; it is a business one too (Campbell 2017). In modern business, collaboration is considered a key component of solving complex problems, and this mentality has shaped the cybersecurity field into a congress of disciplines.

Malicious Cryptography: Exposing Cryptovirology by Adam Young and Moti Yung (2004) is a book that is an extension to their earlier research paper *Cryptovirology*, which is credited with demonstrating the first cryptoviral extortion attack. The book is a compendium of malicious hardware and software attacks geared towards destabilising user and enterprise computer systems. The book was published over a decade prior to cryptolocker attacks in 2013, which was the first major global ransomware attack. Despite this, the authors expertly postulate that in all likelihood, the malware attacks that are described in this book probably represent the tip of the iceberg in terms of what is possible. The research findings are visionary, including successfully identifying key strategic and behavioural patterns of malware developers:

> What the authors of malicious cryptography have done very successfully is to capture the essence of how security can be subverted in this non-standard environment. On several occasions, they refer to game theory without actually invoking the formalism of game theory — emphasizing instead the game-like setting in which security is the value of the ongoing competition between a system designer and its attackers (Young and Yung 2004, p. XV).

The research is innovative and revolutionary that extends beyond the field of computer science. Young and Yung argue that from the perspective of the attack developer, the weakest and most challenging component of the extortion attack is securing the ransom payment without getting caught. For this reason, it may be wise for the attacker to avoid demanding cash entirely (Young and Yung 2004, p. 80). The following example by von Solms and Naccache (1992) is used to further illustrate the problem:

> Consider the problem that a kidnapper faces in collecting the ransom money. Even if the kidnapper sends the delivery person on a wild goose chase to the final drop-off point, the kidnapper may still get caught. This situation is risky from the kidnapper's perspective since it makes certain assumptions about the surveillance capabilities and overall manpower of law enforcement. But with e-money, this is not the case. The kidnapper can insist that the ransom be paid using e-money that is encrypted under the kidnapper's public key. Like in the crypto virus attack, the kidnapper can insist that the encrypted e-money be sent using an anonymous remailer in an anonymous reply. This is known as the perfect crime (von Solms and Naccache 1992).

Building on this argument, Young and Yung identify a key advantage of using a non-fiat currency (e-money) that if the nodes involved spanned multiple nation states, then the tracing effort would be impeded further due to legal hurdles.[7] They hypothesise that "any such development would become a repository for crime syndicates worldwide and would probably amass an endowment in short order that would permit it to invest in legitimate businesses" (Young and Yung 2004, p. 93). Their research also predicted once adopted by the public, that nation states would attempt to pass legislation to prohibit the use of e-money, but cryptography would assist e-money users to circumvent law enforcement's ability to potentially enforce such laws.

From the outset, Young and Yung considered the social impacts of the research. They acknowledge that numerous readers will inevitably object to the nature and public release of this research. However, they argue that "these attacks exist, they are real, and that it is perilous to sweep them under the rug. We believe that ultimately they will surface sooner or later" (Young and Yung 2004, pp. xxiii–xxiv).

---

[7] Note: In 1983, David Chaum described digital money in a scientific paper. A key point that distiguisehs digital money from credit-card payments is anonymity. Users receive the digital currency from their bank, but then it is made anonymous. This allows the bank to see who has exchanged how much money but not what it is used for.

Their research intent was to draw attention to the more serious threats that computer systems face, with the hope that this would encourage the next generation to study cryptography and cybersecurity.

1.3 The Congress of Disciplines

Researchers from multiple disciplines have analysed ransomware attacks and why victims pay ransoms. One of the most influential cyber strategist is Peter Singer who explains that "when a user's computer displays a message threatening to expose activities on pornographic website, fear of exposure can motivate payment" (Singer and Friedman 2014, p. 40). This fear of exposure can often be exacerbated by a user having limited in-depth knowledge of the Internet, cybersecurity or the technology being used. The user may also feel unclear about what is happening, further inducing the fear that the adversary may have done something or planted something illegal (such as child pornography) on their computer that would lead to shame and perhaps prosecution (Lucas 2015, p. 99). These incidents are also likely to occur without being reported to employers or law enforcement, creating widespread under-reporting of cybercrimes.

In an early case of ransomware, Singer details how in 2008 online casinos were threatened with ransomware attacks that were deliberately timed to interrupt their accepting of bets in the final days before the Super Bowl. The attackers demanded the casino pay them $40,000 (USD) to restore the infected systems (Singer and Friedman 2014, p. 88). In these types of targeted attacks, "the victim has to weigh the potential cost of fighting a well-organised attack versus paying off the potential hacker" (Singer and Friedman 2014, p. 88). Organised cybercrime syndicates are not dissimilar to multinational firms; they are seeking maximum return on investments coupled with the lowest operational risk and operating cost. This correlation between organised crime syndicates operations and multinational firms is supported by economics professor Isaac Ehrlich who states:

> The research literature provides abundant evidence that, like multinational firms, organised crime groups consider a number of factors when making decisions related to geographic location of their activities. Perhaps the most important factor influencing the location decision is the strength of the rule of law. A person's decision to participate in an illegal activity is a function of the expected probability of apprehension and conviction and the expected penalty if convicted (Ehrlich 1996).

The risk of attribution, apprehension and conviction may be even further reduced online, with Mittelman and Johnston (1999) stating that "many developing countries' weak rule of law and permissive regulatory regimes provide a fertile ground for criminal activities" (Mittelman and Johnston 1999). Ablon et al. (2014b) reveals how "the hacker market—once a varied landscape of discrete, ad hoc networks of individuals initially motivated by little more than ego and notoriety—has emerged as a playground of financially driven, highly organized, and sophisticated groups" (Ablon et al. 2014b, p. xi).

Josephine Wolff (2018), a researcher from the Harvard's Berkman Klein Center, further extends these strategic risk reduction arguments by explaining that the "evolution of financially motivated cybercrimes, from payment card fraud to ransomware, has been guided in large part by criminals' preference for business models that do not pit them against powerful, centralised intermediaries who can unilaterally monitor or cut off their profits" (Josephine Wolff 2018, p. 77). Empirically, when ransomware attacks have infected large corporations and government systems, these infections have rapidly drawn the interest of law enforcement and the broader cybersecurity communities. Whilst this occurrence may increase the probability of a single organisation paying a large ransom payment, it also increases the resources the organisation and law enforcement can expend on the response and investigation of the attack.

Greenberg (2019) argues that ransomware has always been a plague in the cybersecurity industry, arguing that "when that extortionate hacking goes beyond encrypting files to fully paralyze computers across a company, it represents not just a mere shakedown, but a crippling disruption" (Greenberg 2019). Ransomware is a frank reminder that in an interconnected world, distance does increase or guarantee defence. He concedes that "Every barbarian is already at every gate. And the network of entanglements in that ether, which have unified and elevated the world for the past 25 years, can, over a few hours on a summer day, bring it to a crashing halt" (Greenberg 2019).

Kello (2017) explains "so far, however, the analysis of cyber issues has effectively been ceded to the technologists. Consequently, public perceptions display the following tendencies: a propensity to think of cyber threats as pernicious lines of code – instead of focusing on the human agents who utilize them" (Kello 2017, p. 43). This is supported by Rochester Institute of Technology Professor Samuel McQuade (2006) who explains that:

> Over time, recurring criminal and police innovation cycles have a ratcheting-up effect akin to a civilian arms race. Crime and policing become increasingly complex as a function of increasingly complex tools and/or techniques available in society and employed by criminals, police or security professionals. The result is perpetually complex, technology-enabled crime policing and security management - a never-ending competition in which police and security professionals will, in general, react to criminological innovation (McQuade 2006).

This ongoing problem cycle of law enforcement struggling to adapt and keep pace with cybercriminals is supported by Treverton et al. (2011) who detail that "technological advances; the increasing movement of goods, services, and information; and changing social, economic, and demographic conditions have altered the nature of the crimes, the criminals, and the environment in which criminals and security officers interact" (Treverton et al. 2011). These rapid changes in environment have created a virtual environment that is conducive to ransomware attacks. Bernier and Treurniet (2010) postulate "because of the rate of change in technology and the speed at which actions occur, the challenge lies in our capability to minimise risk and respond appropriately to an attack" (Bernier and Treurniet 2010, p. 238). Acknowledging this environmental change means that we need to implement and further develop dynamic risk assessments, moving beyond the static ones used today. Our situational awareness must match how our networks are being designed and operated today.

1.4 Conclusion

This chapter introduced the genesis of ransomware, highlighting key issues for understanding the successful evolution of ransomware attacks. The speed in which ransomware threat has emerged since 2016 serves reason to why there is a lack of comprehensive in-depth academic literature on ransomware. The chapter emphasises the influence that multidisciplinary scholars have had in the formation and continuous development of cybersecurity as a field, with reference to ransomware. Whilst Gazet's research outcomes are ultimately challenged, they highlight the speed of change in the cybersecurity discipline. Without the advent and adoption of cryptocurrencies, Gazet's research outcomes may have remained accurate today.

Young and Yung's visionary works demonstrated the use of encryption for malicious attacks and successfully predicted that the creation of cryptocurrencies would enable ransomware to evolve into the perfect crime. Schneier's *Applied Cryptography* signalled the beginning of the demilitarisation of encryption to becoming open source. Singer's *Wired for War* and *Cybersecurity and Cyberwar* detailed the rapid speed of emerging technology adoption, whilst Wolff's research challenges cybercrime economics and how we construct modern cybersecurity defences.

The diverse contributions of these scholars highlight the dynamic and complex nature of cybersecurity as an emerging discourse. The congress of disciplines indicates that computer science is not the totality of cybersecurity but the nucleus where computer science began, converging with other disciplines such as economics, politics, mathematics, law, social and behavioural sciences. The literature's composition is diverse, with literary contributions from professional services and managed service providers, which reflects where the front line is in the battle against ransomware attacks. This research unifies and expands on these contributions to create a new multidisciplinary approach to understanding the threat of ransomware.

References

L. Ablon, M. Libicki, A. Golay, Projections and predictions for the black market, in *Markets for Cybercrime Tools and Stolen Data Book Subtitle: Hackers' Bazaar*, (RAND Corporation, Santa Monica, 2014a)

L. Ablon, M. Libicki, A. Golay, Characteristics of the black market, in *Markets for Cybercrime Tools and Stolen Data Book Subtitle: Hackers' Bazaar*, (RAND Corporation, Santa Monica, 2014b)

S. Aftergood, A forum for classified research on cybersecurity, *Fedration of American Scientists*. (2019). Available online: https://fas.org/blogs/secrecy/2018/04/jscore-toc/. Accessed 7 July 2019

M. Albrecht, Ransomware timeline 2010-2017 (2017)

Anonymous, The AIDS Trojan Horse. Comput. Secur. Digest **16**(4), 1–2 (1998)

G. Austin, J. Slay, Australia's response to advanced technology threats: An agenda for the next government, *Australian Centre for Cyber Security*, Discussion paper no. 3 (2016)

M. Bernier, J. Treurniet, Understanding cyber operations in a Canadian strategic context: More than C4ISR, More than Cno, *Conference on Cyber Conflict*. Tallinn, Estonia, 2010: CCD COE Publications

T. Bossert, Press briefing on the attribution of the WannaCry malware attack to North Korea, 19 Dec 2017

N. Campbell, Cyber security is a business risk, not just an IT problem, *Forbes*. (2017). Available online: https://www.forbes.com/sites/edelmantechnology/2017/10/11/cyber-security-is-a-business-risk-not-just-an-it-problem/#3b699de27832. Accessed 13 Nov 2018

A.W. Coburn, J. Daffron, A. Smith, J. Bordeau, É. Leverett, S. Sweeney, T. Harvey, *Cyber Risk Outlook 2018* (Centre for Risk Management Studies, University of Cambridge, in collaboration with Risk Management Solutions, Inc., 2018)

M. DeGusta, Are smart phones spreading faster than any technology in human history? MIT Technol. Rev. (2012). Available online: https://www.technologyreview.com/s/427787/are-smart-phones-spreading-faster-than-any-technology-in-human-history/. Accessed 17 Mar 2019

EastWest Institute, Mobilizing for international action, *Second worldwide cybersecurity summit*. London, 1–2 June 2011 2011 EastWest Institute, 4–5

I. Ehrlich, Crime, punishment, and the market for offenses? J. Econ. Perspect. **10**(1), 43–67 (1996)

A. Gazet, Comparative analysis of various ransomware virii. J. Comput. Virol. **6**(1), 77–90 (2010)

A. Greenberg, A guide to Lockergoga, the ransomware crippling industrial firms, *WIRED*. (2019). Available online: https://www.wired.com/story/lockergoga-ransomware-crippling-industrial-firms/. Accessed 18 Apr 2019

JASON, Science of cyber-security. (2010)

L. Kello, The quest for cyber theory, in *The Virtual Weapon and International Order*, (Yale University Press, 2017)

A. Knott, Towards fundamental science of cyber security, in *Network Science and Cybersecurity*, (Springer, 2014), pp. 1–13

E. Lucas, *Cyberphobia: Identity, Trust, Security and the Internet* (Bloomsbury Publishing, London, 2015)

S. McQuade, Technology-enabled Crime, Policing and Security. J. Technol. Stud. **32**(1/2), 32–42 (2006)

J. Mittelman, R. Johnston, The globalization of organized crime, the courtesan state, and the corruption of civil society. Glob. Gov. **5**(1), 103–126 (1999)

T. Moore, E. Kenneally, M. Collett, P. Thapa, Valuing cybersecurity research datasets, *Workshop on the Economics of Information Security (WEIS)*. Cambridge, MA, June 3–4, 2019, The University of Tulsa, International Computer Science Institute, Berkeley Office of Science & Technology, and Department of Homeland Security

P. Mungo, B. Clough, *Approaching Zero: The Extraordinary Underworld of Hackers, Phreakers, Virus Writers, and Keyboard Criminals* (Random House, New York, 1992)

D. Nazarov, O. Emelyanova, Blackmailer: The story of Gpcode, *Securelist*. (2006). Available online: https://securelist.com/blackmailer-the-story-of-gpcode/36089/. Accessed 10 Jan 2019

H. Orman, Evil Offspring – Ransomware and crypto technology. *IEEE Internet Comput.* **20**, 5 (2016)

A. Simone, The strange history of ransomware: Floppy disks, AIDS research, and a Panama P.O. Box., *Medium*, (2015). Available online: https://medium.com/un-hackable/the-bizarre-pre-internet-history-of-ransomware-bb480a652b4b. Accessed 9 Sept 2018

B. Sims, 'NotPetya' ransomware attack spreading rapidly to organisations across the globe, *RiskXtra*, (2017). Available online: https://www.risk-uk.com/notpetya-ransomware-attack-spreading-rapidly-organisations-across-globe/. Accessed 3 Jan 2019

P. Singer, A. Friedman, *Cybersecurity and Cyberwar: What Everyone Needs to Know* (Oxford University Press, New York, 2014)

A. Solomon, B. Nielson, S. Meldrum, AIDS technical information, *The Center for Education and Research in Information Assurance and Security*. (2000). Available online: http://ftp.cerias.purdue.edu/pub/doc/general/aids.tech.info. Accessed 10 Sept 2017

A. Taylor, *NotPetya Malware Attributed*. (16 Feb 2018)

D. Terr, Polynomial Tim, *MathWorld.* (2019). Available online: http://mathworld.wolfram.com/PolynomialTime.html. Accessed 13 Jan 2019

G. Treverton, M. Wollman, E. Wilke, D. Lai, The threat will continue to morph, in *Moving Toward the Future of Policing*, (RAND Corporation, 2011), pp. 89–106

S. von Solms, D. Naccache, On blind signatures and perfect crimes. *Comput. Secur.* **11**(6), 581–583 (1992)

J. Wolff, *You'll See This Message When It Is Too Late: The Legal and Economic Aftermath of Cybersecurity Breaches* (The MIT Press, Cambridge, 2018)

A. Young, M. Yung, Cryptovirology: Extortion-based security threats and countermeasures, in *Proceedings 1996 IEEE Symposium on Security and Privacy*. 6–8 May 1996

A. Young, M. Yung, *Malicious Cryptography: Exposing Cryptovirology* (Wiley Publishing, Indianapolis, 2004)

Chapter 2
Genesis of Ransomware

This chapter discusses the formation of ransomware attacks, adaptive attack methodologies and how ransomware attacks can be classified. Analysis of the major ransomware attacks highlights why these cyberattacks have and continue to pose such a significant threat to critical infrastructure, governments, enterprises and individual citizens' devices and networks. The following section of the chapter discusses common ransomware countermeasures, their limitations and alternative approaches to prevent and detect ransomware attacks. The final section briefly details some of the major ransomware attacks that have occurred and the rapid increase in attack volume and encryption sophistication since 2013.

Since the turn of the twenty-first century, cybercrime has become organised and industrialised like no other crime (Moore et al. 2009). This maturing process has enabled would-be cybercriminals to continuously hone their skills whilst searching for the next opportunity to profit. In turn, this has created numerous cybersecurity challenges for enterprise cyber risk managers and security operations teams. Despite the continued rise of numerous new threats, it is ransomware that is now the most dangerous form of cyberattack that enterprises must be prepared for. Greenberg (2019) explains that a ransomware attack can rapidly cripple an organisations ability to operate, applying pressure for every minute lost, and that is costing that enterprise a significant amount of money (Greenberg 2019). In the wake of rapid technological advances in applied cryptography, the Internet and financial systems, enterprises are now more vulnerable to ransomware attacks than ever before.

M. Ryan, *Ransomware Revolution: The Rise of a Prodigious Cyber Threat*,
Advances in Information Security 85, https://doi.org/10.1007/978-3-030-66583-8_2

2.1 Ransomware Taxonomy

Cyberattacks have historically been categorised by their method of delivery and desired impact. Ransomware attacks may disrupt or deny a user's ability to access data or operate a device or network; therefore, this form of cyberattack has commonly been categorised as an extortion or data integrity attack, not a denial-of-service attack. Ransomware attacks have typically delivered their malicious payload through automated scripts and links embedded within emails, data files and websites. The modus operandi can vary between attackers; however, the most common characteristic of ransomware stems from a piece of malicious software being inadvertently run by the user, resulting in the user's system files becoming encrypted. Ransomware attacks are designed to render the users' data unusable; commonly, folders become locked (encrypted), and pictures and text-related files become scrambled bodies of text (National Institute of Standards and Technology 2018). Table 2.1 below provides an illustrative demonstration of the data encryption process.

Another characteristic of ransomware is the victim notification process. Ransomware attacks are generally designed to be executed in the shadows of the operating system and network until the encryption process has successfully been completed. Once a user's system has been compromised (infected), the malicious code informs the user of their impending situation. Typically, the user's display is commandeered which alerts the user to the presence of the ransomware attack, which also enables the attacker to make a direct request to the victim about the required ransom payment. The request for payment prompt (message) will commonly inform the victim of their predicament whilst also providing details of how to make the payment and the required time frame for payment. The attacker will typically demand the payment be made through an untraceable medium, such as untraceable gift cards, or more commonly through a cryptocurrency such as Bitcoin, Ripple or Monero. In a trend akin to many telecommunications scams, some requests for payment claim to be sent by law enforcement agencies as detailed below in Fig. 2.1.

Table 2.1 Plain text comparison with cipher text

Plain text	Cipher text (AES 256)
Applied cryptography (encryption) is at the heart of today's financial and military communication and storage systems. The same algorithms that are used to protect our sensitive data at rest and during transmission are fundamental components of ransomware attacks.	Z9aw3U08bzuDCfgFEQuurvmboonjqD6K+uf/mPRarUBo7WVTLzqc9Twi4q5oAgDXHktZZaGdFzWh/AJ21pMmQO42XJFEsa9rqp0Rp8n6jH/cBa7qEWxwb9dq4zT0wGXSgX61kBhcfbYsome6uLO0RlIEDQQPDSCt6vVwiSZKP47U7VuOaUbAY+08qytJyRh0i5k3+MH5kPxwHCKFeR6oVRETA8Q/FGhNd5sl5JBsPKONAM731PxwFeQTMoIAAZ05zEwW69XAgBBODXDC5EMYBq9KB4PfO7q56z9LYxWnXJ276wewb47nDg+bwKjVKAaRYSPOPbqfp+7CVBX9jSaGyM0S+Cj87Y/bX4Jv3cjcufb4jYIoTMf4dDBKanFcxGUD

Fig. 2.1 Sample ransomware request for payment

In 2014, this request for payment approach was utilised throughout the Virlock ransomware attacks. The Virlock attacks quickly spread around the globe, infecting victims primarily located in the United States, China and Australia. One of the reasons Virlock was so difficult to detect and defend against was its adaptive coding structure. Virlock was designed using polymorphic code, a scenario where the code changes (mutates) every time it is run, without altering the primary function of the code (Richet 2015). The mutating code was used to conceal the malicious code in the target's operating systems and to randomly generate a different encryption key for each infected host. This random key generation prevented one single key being used to eliminate or decrypt multiple infected hosts.

Since the emergence of ransomware, the malicious code underpinning ransomware has continued to evolve in complexity, creativity and discretionary targeting. Many computer scientists and cybersecurity experts anticipate that the spread of ransomware will continue to grow and spread through new devices and platforms. A multitude of experts have also argued that what works on personal computers can easily be adapted for use in mobile device environments (Becher et al. 2011). The spread of ransomware to Internet of Things (IoT) devices could pose a significant cybersecurity threat because IoT devices do not require or facilitate any form of security patching or updates.

Due to the relatively low cost of IoT devices, should a device become infected, it may not be financially viable to take any course of action other than paying the ransom or disposing of the device. This could lead to potentially billions of devices being rendered useless from a single ransomware attack. In the foreseeable future, this may have broader implications beyond popular low-cost consumer IoT devices such as consumer electronics, sensors, cameras and smartwatches. In more complex environments, Zhang et al. (2014) raised security concerns that:

> Ransomware could allow an attacker to remotely disable selected vehicle functions (e.g., lock the doors or the in-car radio, immobilize the engine) in a way that the vehicle owner's car keys can no longer activate them. The attackers can then demand ransom to be paid before re-enabling these functions (Zhang et al. 2014, p. 14).

These serious security concerns are not unwarranted. In January 2018, one of Schneider Electric's critical infrastructure clients was targeted by a cyberattack dubbed Triton. Schneider Electric is one of the world's largest producers of industrial hardware and software for the energy management and automation solutions industry. The Triton attack targeted one of Schneider's critical infrastructure customers initially by exploiting a previously unknown vulnerability (zero day)[1] against Schneider's Triconex Tricon safety system's firmware. In the second stage of the attack, the attackers deployed a remote access Trojan, enabling the attacker to remotely execute commands. This is the first known occurrence of malware targeting industrial control system using this methodology (Newman 2018). Schneider believes the attacker was ultimately trying to manipulate layers of the built-in emergency shut down protocols. This type of cyberattack highlights one of the significant threat ransomware poses the broader community, and owners and operators of critical infrastructure should anticipate and be prepared for ransomware attacks.

The attack on Schneider Electric's embodies a growing concern about the cybersecurity and safety of industrial and critical infrastructure sites. These concerns are highlighted by emerging research by Al-Hawawreh et al. (2019) into the Industrial Internet of Things (IIoT), which found "much of the commercial value of such systems resides at their edge tier, where significant storing, analysing, processing, and controlling capabilities are located. This makes edge systems an attractive target for advanced security threats such as accelerating extortion attacks, which include ransomware" (AL-Hawawreh et al. 2019). The research also suggests that due to the ease of access and functionality, IIoT edge gateways may become the preferred target for deploying ransomware attacks.

2.2 Classification

From a technical analysis perspective, ransomware attacks are more broadly classified as a type of malware. Malware is a commonly used term to describe a piece of malicious software or file that is intentionally designed to be harmful to a computer

[1] Note: A zero-day exploit is the term used to describe a bug or vulnerability within a software or hardware platform that has not been reported to the product developer or to its users. See Ganame et al. (2017).

or electronic device. Malware is typically designed to be a pre-packaged exploitation of a known or unknown vulnerability. Once activated, the malware is designed to deliver a "payload" of actions and instructions. The instructions can include details about what the system should do after it has been compromised. The activation mechanism can vary between requiring the user to click a link or enable hidden content, or it may be remotely activated by the attacker. Some types of malware also employ additional functions such as worms, which contain a set of instructions that enable an automated self-reproduction cycle to rapidly propagate the attack (Singer and Friedman 2014, p. 40).

Analysis of ransomware attacks and source code indicates that ransomware attacks could be further categorised as targeted or non-targeted attacks. Targeted attacks occur when the attacker develops a ransomware attack for a specific target. The reasoning behind the target selection can range from being purely financially motivated, revenge, geographical location or simply because the individual or organisation may be using a vulnerable application or device (Singer and Friedman 2014, p. 88). The language used may also be another indicator of the desired target and the origin of the attacker. Research by TrendMicro in 2017 indicated the overwhelming majority of reported ransomware attacks occurred in four major geographical regions. The Asia Pacific region was the most targeted with 33 percent, followed by Europe and the Middle East with 25 percent, then Latin and South American region 23 percent and 15 percent in the North American region (Micro 2017b).

Alternatively, non-targeted attacks are indiscriminate and have no predefined infection limitations. These types of ransomware attacks typically utilise a worm or similar propagation mechanism to rapidly spread the infection through large numbers of user systems and devices. However, despite being classified as non-targeted ransomware, this may have limited impact on victim geography versus targeted attacks. This is because the largest number of consumer devices, networks and IoT are operated in these areas; thus, the attack surface is much larger due to the volume of devices.

2.3 Situation Evaluation

When an individual or organisation falls victim to a ransomware attack, and they have not implemented an effective business continuity plan (BCP), it is imperative to understand the odds of a successful recovery from the outset. Without backups or an effective BCP, ransomware victims are confronted with the options of forgoing their data and starting again or attempting to break the attacks' encryption. Technical investigations have successfully recovered victim's decryption keys through flawed software development. Outside of the attacker making an error during the ransomware's source code development, the remaining option is to crack the decryption key. Encryption is a puzzle created through mathematics, so it makes sense to use mathematics to solve the puzzle.[2] To assist in this process and provide users a broader perspective, this research considered some of the fastest known computers in operation today.

Whilst it is difficult to ascertain the true computational power of advanced military superpowers, numerous articles have emerged in recent years describing some of these pursuits. One of the first known projects was named Jaguar for its speed, and it was rumoured to be capable of executing "a quadrillion (10^{15}) operations per second (known as a petaflop). In 2009, the Jaguar supercomputer was clocked at 1.75 petaflops, officially becoming the world's fastest computer" (Singleton 2018). Shortly after this, the NSA began development on their next supercomputer codenamed Titan, whose objective is to achieve speeds of up to 20 petaflops by 2018. The United States is not the only state developing supercomputers, with numerous projects in operation in China, Japan and the United Kingdom. In November 2017, China's Sunway "TaihuLight" peaked at a speed of 93 petaflops per second. Seven months later, the Department of Energy and IBM in June announced that America's latest supercomputer dubbed "Summit" had achieved speeds of 200 petaflops per second, more than doubling China's Sunway TaihuLight (Singleton 2018).

It is easy to cite these extraordinary computational speeds; however, in reality, the amount of resources and expertise required to achieve these supercomputer computational speeds is immense. For instance, to undertake project Titan, the NSA needed to construct a building 260,000 square feet that required 200 megawatts of power to operate the required equipment. This is enough power to provide power to a small city with 200,000 homes. In another similar project for the US National Energy Research Scientific Computing Center (NERSC), the cost to deliver the project's Cray XC30 supercomputer was $70 million (USD). Despite these phenomenal speeds and resources required, none of these supercomputers could come close to decrypting all possible AES 256-bit keys in under a year. This is greater than 100 times the duration of most ransomware request for payment windows.

Another theoretical method to examine the effectiveness of an encryption standard is to review the relevant recommended intelligence and defence agency encryption standards. For example, within Australia if ASIO were aware or capable of easily breaking AES 128-bit or AES 256-bit encryption, they would remove it from their recommendations and standards. In 2019, the Australian Signals Directorate (ASD) was recommending defence and intelligence agencies to use AES key lengths of 128, 192 and 256 bits (Australian Signals Directorate 2020). Whilst Australia is not a cyber superpower, the Five Eyes alliance ensures Australian strategic agencies must have a minimum level of competency for handling sensitive information.

So what does this all mean for an individual or corporation who has become a victim from a ransomware attack? It means that even with access to a next-generation supercomputer, it is mathematically unlikely the victim will be able to discover the

[2] Note: Whilst decryption can potentially be sped up further by the application of decryption tools such as password dictionaries and rainbow tables, the decryption process remains a mathematical process that ultimately may or may not be significantly influenced by the application of these decryption tools.

encryption key using brute-force techniques. Whilst the attacker may have made a mistake by using a flawed encryption process or easy-to-break key, the statistical odds are not in the victim's favour to successfully brute force the encryption key within typically defined payment deadlines.

2.4 Recovery Times

Recovery times for individuals and organisations that fall victim to ransomware attacks are subject to a variety of factors which create inconsistent recovery time outcomes. In the simplest scenario, victims paid the ransom demands and were promptly provided the key to decrypt their data, enabling rapid restoration of operations on their networks and devices. On the other end of the spectrum, victims paid the ransom demands and never received the decryption keys or any further correspondence from the attacker. Technical investigations by security researchers and vendors have also discovered that some ransomware attacks were designed in a manner (i.e., key deletion) that prevented recovery, even if the ransom was to be paid.

Data analysis of ransomware attacks in 2017 by the Ponemon Institute indicates on average, it took organisations 23 days to resolve and recover from a ransomware attack (Ponemon Institute 2017, p. 9). An adjacent ransomware study by Symantec during the same period revealed the average ransomware demand was $522 (USD) (Symantec Security Center 2019). Further information can be derived by analysing these results. Given that the average ransomware demand window is between 48 and 96 hours, these results suggest that many victims never paid, or their data was never recovered. Whilst it makes sense from a marketing perspective to decrypt victim's data when ransoms are paid, this is not always the case.[3] Victims of ransomware can be confronted with a complex set of problems that have an inherently high degree of uncertainty.

Another contributing factor is that many large corporations may not be in a position to pay ransom demands in cryptocurrencies. Whilst exchanges between fiat and cryptocurrencies continue to emerge and become more mainstream, few large corporations actively trade and undertake day-to-day business using cryptocurrencies. This creates an additional layer of complexity for decision makers and executive boards. As a result, many large corporations may elect to not pay ransom demands because it is simply too hard, instead choosing to direct all their available resources to alternative recovery processes. Research undertaken by the Ponemon

[3] Note: Whilst there are exceptions, cybercriminals are a business, and from that perspective when they threaten to leak a victims data, they must make good on that threat – their business model is reliant on the element of fear. In the same way, when their ransomware attack spreads, they are reliant on victims informing the next victim just to pay the ransom demand, and they will get their data back. If profit is the primary objective, then the attackers must remove any doubt from the victim's mind in order to get paid.

Institute indicated that of the 254 large organisations surveyed, 98 percent incurred at least one malware attack in 2016 and 2017. The research also identified that the number of ransomware attacks doubled in 2017 versus 2016, indicating that one in ten had incurred a ransomware attack in 2017 (Ponemon Institute 2017).

2.5 Ransomware Countermeasures

Despite having open access to a taxonomy containing hundreds of different security and risk management controls, corporations remain unprepared to respond to ransomware attacks (United States Department of Commerce 2019). An inherent problem with most cyber risk management strategies and cybersecurity controls is their static nature; on the contrary, ransomware attacks are constantly evolving and dynamic.[4] However, the evolution of attacks does not indicate that defenders are constantly engaged in a high-stakes game of brinkmanship, where an attacker's success is based on their ability to continuously develop new vulnerabilities. In fact, many targets fall victim to ransomware attacks that used already known vulnerabilities which had effective and free countermeasures readily available.

Analysis of major ransomware attacks indicates attack success is commonly achieved through two distinctive time-orientated pathways. The first is through the development and deployment of zero-day exploits whereby the attacker will use a previously unknown attack vector to infiltrate and corrupt a given network or system. This type of attack presents the greatest cyberthreat to corporations and governments because cyber defence strategies and tools are predicated on known attack vectors and methods. Philosophically, one cannot defend an attack vector that one does not know exists. The existence and continued development of zero-day exploits by government agencies and criminal organisations presents an ongoing threat to global cybersecurity. From an attacker's perspective, even when new ransomware attacks are successfully thwarted by defenders, attackers can draw on the nefarious Irish Republican Army (IRA) statement, "You have to be lucky all the time. We only have to be lucky once" (Clutterbuck 2019).

The second pathway is through what can be best described as viral latency or progeny whereby a known ransomware strain is re-triggered, re-engineered or redeployed using known vulnerabilities. This pathway naturally raises the following question: If a vulnerability is well known, and a solution is readily available for deployment, why do these types of ransomware attacks remain so successful? The answers to this problem are diverse and range in complexity, and understanding how security controls are generally deployed in different environments is an important component of trying to further understand the success of ransomware attacks.

---

[4] Note: Many organisations continue to assess their cyber risk periodical basis (i.e., quarterly or annually). Whilst organisations are increasingly monitoring their external environments, the processes required to alter their risk profiles and act on this information remain relatively underdeveloped.

2.5.1 Deploying Countermeasures

For corporations and governments, the deployment of cybersecurity defences and controls is generally established through the creation of a variety of attack scenarios which are input into a series of risk management processes.[5] However, Wolff successfully argues that high-level frameworks for information security like the Confidentiality Integrity Availability (CIA) triad cannot be easily translated into an operational understanding of how to implement security controls, so too that operational understanding laid out in NIST and ISO catalogues cannot be easily mapped back to a coherent, consistent high-level cybersecurity framework. There is a fundamental disconnect between the literature describing high-level information security frameworks and low-level information security controls (Wolff 2015, p. 48).

This is supported by Stolfo et al. (2011) who state that "in recent years, much research has provided a strong understanding of a particular vulnerability or the security issues involved in designing a given system. However, we've seen little success in collecting the knowledge from this research into a general, systematic framework" (Stolfo et al. 2011). One goal is to establish a common framework for classes of defences, which can be categorised based on the policies they can enforce and the types of attacks they can potentially prevent or detect. Richard Bejtlich (2013) outlines that within cybersecurity practice, there is an underlying acknowledgement that determined adversaries will inevitably breach your defences (Bejtlich 2013, p. 5). This mindset is present consistently within providers of security services, and organisations should develop security strategies that not only prevent but also anticipate breaches occurring.

2.5.2 Effective Countermeasures

Most specific anti-ransomware applications focus on "analysing real-time log alerts generated by file activity monitoring tools. These tools aim to identify suspicious rapid and numerous file changes reflecting unapproved data deletion or modification such as encryption" (Australian Cyber Security Centre 2018). Sir Rob Wainwright (2019), a partner at Deloitte Netherlands, explains that "companies have adopted a range of tactical and strategic security measures to counter the threats they face in cyberspace. However, none provide 100 percent protection" (Wainwright 2019). Therefore, it is inevitable that attacks will be successful, and breaches will occur. The precipitous increase in cyberattack volumes, complexity and seriousness has forced cybersecurity to become a serious concern at the board level.

---

[5] Note: For example, organisations may elect to use generic ISO risk management standards or more specialised cybersecurity risk management framework such as NIST 800, ISO27001, COBIT5, or a combination of multiple standards.

Security vendor TrendMicro advises their clients that as ransomware further evolves, so should the IT/system administrators and information security professionals who protect their organisation's crown jewels; end users must be equally proactive, and defence-in-depth must be practiced to combat ransomware attacks (Micro 2017a). However, the use of the term defence-in-depth can be problematic in cybersecurity because it is not well understood. Wolff explains that:

> While military historians and nuclear safety experts have defined accepted, field-specific notions of defence-in-depth, computer security specialists have instead seized on the term and invoked it so often, so inconsistently, and in such vague terms that it generally amounts to little more than a reassuring label for lots of defence or more than one defence (Wolff 2015, p. 36).

Within literature on military strategy, defence-in-depth approaches are not focused on adding more defences to a kingdom but rather about reconfiguring and diversifying defensive resources to adjust the terms of victory and defeat (Wolff 2018). This strategic objective is often absent from cybersecurity approaches, which have attempted to achieve defence-in-depth through simply deploying additional controls.

Whilst cybersecurity controls and best practices reduce the likelihood of becoming the next ransomware victim, there is no silver bullet. In 2019, Microsoft reported there is a trend towards increasingly sophisticated ransomware behaviour. Large organisations continue to be considered high-value targets, with older platforms especially susceptible to ransomware attacks. To reduce the threat posed by ransomware, Microsoft recommends users:

> Back up important files regularly. Use the 3-2-1 rule. Keep three backups of your data, on two different storage types, and at least one backup offsite. Apply the latest security updates to your operating systems and applications. Educate your employees so they can identify social engineering and spear-phishing attacks. Controlled folder access. It can stop ransomware from encrypting files and holding the files for ransom (Levin and Simpson 2019).

This simplistic approach highlights the diversity and complexity of preventing ransomware attacks. From an NIST 800 perspective, the first recommendation is a resilience control measure, the second is a tactical (ad hoc) solution, the third is education and awareness control, and the final is a technical control that is achieved through design. Whilst these recommendations are basic in nature, implementing even these basic controls can present complex challenges for large organisations.

Essential cybersecurity practices such as patching and vulnerability management may cause only a few minutes of downtime for a home user to install and implement the latest updates, but for a corporation, this may be an expensive and resource-exhausting process. Many large corporations have in excess of a million devices in their IT and OT environment. Blindly rolling out patches may require those systems to be down for a matter of minutes, but that process may trigger a myriad of downstream issues which result in losses in customer services, productivity and ultimately financial losses.

To avoid outages, many corporations undertake extensive testing processes before planning to roll out new patches. This is a time-consuming process and results in a slow but steady gradual roll out of patches across the IT environment. Another obstacle is large organisations may have thousands of different applications that require different patches. Additionally, this problem is made more challenging because most large corporations outsource portions of their IT infrastructure management to multiple external service providers. As a result, corporations may have millions of devices that require different security patches across internally and externally controlled environments. With finite time and resources, this is a problem that for many cannot be solved.

Even when organisations follow and implement industry best practices such as robust backups, they may not escape the threat of ransomware. The following example from the Australian Cyber Security Centre highlights the disconnect between industry best practices and the practical applications of security controls:

> In 2016, an Australian government organisation identified ransomware on a user computer and responded by simply reimaging the computer's hard drive. Three months later, the organisation's IT staff realised that thousands of files needed for legal proceedings and stored on a network drive (file share) had also been encrypted by the ransomware. Due to the amount of time that had elapsed, the organisation's backups contained encrypted copies of the files (Australian Cyber Security Centre 2018, p. 37).

This simple example demonstrates the complexity organisations face in trying to prevent and encounter responding to ransomware attacks. Analysis of ransomware security best practices indicates there is prevention advice openly available; however, there are differences in the recommended approaches. This also highlights the lack of a universal solution to preventing ransomware attacks. Cybersecurity is an exhaustive process for large organisations, and it requires continuous evaluation and improvement.

2.5.3 Alternative Approaches to Countering Ransomware

Security providers such as Kaspersky have previously had success in reverse-engineering some ransomware attacks to develop decryptors to help victims recover from attacks; however, this process is dependent on the quality of encryption and source code used by the attackers. This approach is countered by Wolff's argument that insists:

> Unlike the soldiers of the Roman Empire, defenders of computer systems do not have a clear geographic perimeter or central, physical capital to focus their protection efforts on, nor a well-defined set of sequential signals of escalating harm, like the protectors of nuclear plants. They have elements of all these things, in the context of particular incidents, or even particular stages of incidents, but there is no single consistent metaphor that can be applied across the range of security threats that involve computers. Instead, we end up with lots of metaphors, mixed and applied haphazardly, and too little sense of what we can actually learn from historical endeavours—and which versions of history we have conveniently invented to explain and reinforce our own ideas (Wolff 2015, p. 173).

Additionally, an inherent problem with using reverse engineering to counter ransomware is the potential time delay. Whilst reverse engineering ransomware attacks has successfully limited the propagation and success of some major ransomware attacks, it is typically a time-consuming and resource-exhaustive process. The volume of ransomware attacks also outpaces the resources available to reverse-engineer them; consequentially, these efforts are limited to the largest attacks.

2.5.4 Alternative Approaches to Ransomware Detection

Emerging electrical engineering research from the US Naval Postgraduate School proposed using an algorithm to detect the encryption process, which is a universal characteristic of all crypto ransomwares. The method proposed by Melton (2018) applies power analysis techniques to classify the steps of encryption and correlate them to identify when encryption is occurring on a computer.

> Due to the nature of the malware, encryption processes are detected. When such processes are observed, it is possible to stop the computer processes until verified by the user. As a result, ransomware activity can be halted before continuing to encrypt data and possibly preventing further damage in data loss. Stopping encryption early in the ransomware's process limits the amount of data being no longer accessible by the user while saving most of the files on a computer (Melton 2018).

From a practical implementation perspective, the research is still in its infancy phases with limited applications and configurations being successfully tested at this point. However, the research produced demonstrates a potential new method to detect ransomware attacks that may be cost-effective to implement across enterprises whilst also providing an additional layer of protection beyond conventional cybersecurity protections such as firewalls and antivirus.

2.6 Major Ransomware Attacks

Many of the largest ransomware attacks in history, their discovery date and the types of encryption algorithms used are detailed in Table 2.2 below. The data indicates that ransomware developers commonly use the same well-known encryption algorithms that governments and organisations trust to keep their valuable data secure. The data also highlights that ransomware developers are duplicating proven encryption processes, with the majority of their efforts being spent on devising new infiltration and infection techniques.

Table 2.2 Major ransomware attack encryption algorithms (Palisse et al. 2016)

Malware	First known occurrence	Encryption algorithms
GPcoder	2004	AES – ECB
Cryptolocker (Gameover ZeuS)	2013	AES
Crypto Wall	2014	AES – CBC
CTB Locker	2014	AES – ECB
Torrent Locker	2014	AES – CTR I CBC
Tesla Crypt	2015	AES – ECB I CBC
Crypt Vault	2015	RSA – OAEP
Locky	2016	AES – CTR I EBC AES RSA + ECB
Petya	2016	Salsa20
NotPetya	2016	MFT – Salsa20 – AES
WannaCry	2016	AES – RSA
SamSam	2016	RSA
Hermes	2017	RSA – AES – CBC
Ryuk	2018	RSA – AES – CBC
GandCrab	2019	RSA – AES I CBC RSA-Salsa20

2.7 Conclusion

This chapter discussed ransomware's taxonomy, formation and the differences between ransomware and alternative forms of cyberattack. The chapter details the adaptive and complex nature of ransomware attacks whilst highlighting the criminal ingenuity of attackers to coerce their victims to pay the ransoms in increasingly anonymous formats. The section also raised security concerns about the design vulnerabilities of IoT devices and their potential exposure to ransomware attacks. These concerns come at a time when the number of uses for IoT devices is exponentially growing, whilst ransomware attacks are simultaneously demonstrating a rapid increase in attack volume and complexity.

Another critical security concern raised was zero-day attacks, which have the potential to circumvent all known security control measures and tools. The deployment of ransomware attacks using zero-day exploits represents a prodigious threat to organisations and government systems. The chapter also questions the effectiveness of high-level cybersecurity frameworks with low-level and operational security controls. The examples demonstrate the complexity organisations face in trying to prevent and encounter responding to ransomware attacks. With no agreed or universal approach to preventing ransomware, organisations will continue to be challenged by ransomware. The final section of the chapter provided insight into a potential ransomware countermeasure that applies power analysis techniques to identify and alert the user to activation of the system's encryption process.

References

M. AL-Hawawreh, F. den Hartog, E. Sitnikova, Targeted ransomware: A new cyber threat to edge system of brownfield industrial Internet of Things. IEEE Internet Things J. **6**(4), 7137–7151 (2019). https://doi.org/10.1109/JIOT.2019.2914390

Australian Signals Directorate, *Australian Government Information Security Manual* (Department of Defence, Canberra, 2020) Available online: https://www.cyber.gov.au/sites/default/files/2019-08/Australian%20Government%20Information%20Security%20Manual%20%28August%202019%29.pdf. Accessed 11 Aug 2019

Australian Cyber Security Centre, *Strategies to Mitigate Cyber Security Incidents – Mitigation Details* (Australian Signals Directorate, Canberra, 2018). Available online: https://www.cyber.gov.au/sites/default/files/2019-03/Mitigation_Strategies_2017_Details_0.pdf. Accessed 23 May 2018

M. Becher, F.C. Freiling, J. Hoffmann, T. Holz, S. Uellenbeck, C. Wolf, Mobile security catching up? revealing the nuts and bolts of the security of mobile devices, *IEEE symposium on security and privacy (SP)*. Oakland, California, USA, 2011, 96–111

R. Bejtlich, *The Practice of Network Security Monitoring: Understanding Incident Detection and Response* (No Starch Press, San Francisco, 2013)

L. Clutterbuck, Terrorists have to be lucky once; targets, every time. Available online: https://www.rand.org/blog/2008/11/terrorists-have-to-be-lucky-once-targets-every-time.html. Accessed 12 May 2019

K. Ganame, M. Allaire, G. Zagdene, O. Boudar, Network behavioral analysis for zero-day malware detection – a case study, in *First International Conference, ISDDC*, (Springer, Vancouver, 2017)

A. Greenberg, A guide to Lockergoga, the ransomware crippling industrial firms, *WIRED*. (2019). Available online: https://www.wired.com/story/lockergoga-ransomware-crippling-industrial-firms/. Accessed 18 Apr 2019

B. Levin, D. Simpson, Ransomware. 4 Apr 2019. Available online: https://docs.microsoft.com/en-us/windows/security/threat-protection/intelligence/ransomware-malware. Accessed 9 May 2019

J. Melton, *Detecting ransomware through power analysis*. Master of Science Electrical Engineering Naval Postgraduate School. (June 2018). Available online: https://calhoun.nps.edu/bitstream/handle/10945/59721/18Jun_Melton_Jacob.pdf?sequence=1&isAllowed=y. Accessed 11 Feb 2019

T. Micro, Best practices: Ransomware. (2017a). Available online: https://www.trendmicro.com/vinfo/us/security/news/cybercrime-and-digital-threats/best-practices-ransomware. Accessed 5 May 2019

T. Micro, Ransomware. (2017b). Available online: https://www.trendmicro.com/vinfo/us/security/definition/ransomware. Accessed 18 May 2019

T. Moore, R. Clayton, R. Anderson, The economics of online crime. J. Econ. Perspect. **23**(3), 3–20 (2009)

National Institute of Standards and Technology, Data integrity: Recovering from ransomware and other destructive events. 1800-11. National Institute of Standards and Technology. (2018). Available online: https://www.nccoe.nist.gov/publication/1800-11/index.html. Accessed 27 Apr 2018

L.H. Newman, Menancing malware shows the dangers of industrial sabotage, *WIRED*. (2018). Available online: https://www.wired.com/story/triton-malware-dangers-industrial-system-sabotage/. Accessed 23 Feb 2018

Palisse, A., H. Le Bouder, J.-L. Lanet, C. Le Guernic, A. Legay, Ransomware and the Legacy Crypto API, *The 11th International Conference on Risks and Security of Internet and Systems*. Roscoff, France, 5th–7th September 2016 (Springer, 2016)

Ponemon Institute, 2017 cost of cyber crime: Insights on the security investments that make a difference, (2017). Available online: https://www.accenture.com/t20170926T072837Z__w__/us-en/_acnmedia/PDF-61/Accenture-2017-CostCyberCrimeStudy.pdf. Accessed 23 May 2018

J.-L. Richet, Extortion on the Internet: The rise of Crypto-ransomware, *Cybercrime*. (2015). Available online: https://blogs.harvard.edu/jeanlouprichet/files/2015/07/Extortion_on_the_Internet_Rise_of_Crypto_Ransomware.pdf. Accessed 1 Aug 2019

P. Singer, A. Friedman, *Cybersecurity and Cyberwar: What Everyone Needs to Know* (Oxford University Press, New York, 2014)

M. Singleton, The World's Fastest Supercomputer is back in America, *The Verge*. 12 June 2018 (2018) [Online]. Available online: https://www.theverge.com/circuitbreaker/2018/6/12/17453918/ibm-summit-worlds-fastest-supercomputer-america-department-of-energy. Accessed 18 Feb 2019

S. Stolfo, S. Bellovin, D. Evans, Measuring security. *IEEE Secur. Priv.* **9**(3), 88 (2011)

Symantec Security Center, What is ransomware? And how to help prevent it. (2019). Available online: https://us.norton.com/internetsecurity-malware-ransomware-5-dos-and-donts.html. Accessed 13 Jan 2020

United States Department of Commerce, *National Institute of Standards and Technology Special Publication 800-53: Security and Privacy Controls for Information Systems and Organizations*. National Institute of Standards and Technology. Available online: https://csrc.nist.gov/csrc/media/publications/sp/800-53/rev-5/draft/documents/sp800-53r5-draft.pdf. Accessed 3 May 2019

R. Wainwright, The ascent of the CISO, *Deloitte Cyber*. (2019). Available online: https://www2.deloitte.com/nl/nl/pages/risk/articles/the-ascent-of-the-ciso.html. Accessed 7 May 2019

J. Wolff, *Classes of Defense for Computer Systems*. Doctor of Philosophy in Engineering Systems: Technology, Management, and Policy Massachusetts Institute of Technology, June 2015

J. Wolff, *You'll See This Message When It Is Too Late: The Legal and Economic Aftermath of Cybersecurity Breaches* (The MIT Press, Cambridge, 2018)

T. Zhang, H. Antunes, S. Aggarwal, Defending connected vehicles against malware: Challenges and a solution framework. *IEEE Internet Things J.* **1**(1), 10–21 (2014)

Chapter 3
Evolution of Applied Cryptography

This chapter briefly explains the evolution of applied cryptography by reflecting on its origins and its ongoing evolution. The objective of this chapter is to provide a basic insight into how applied cryptography has evolved whilst acknowledging the contributions of the inventors and security researchers that have further advanced the discipline. Indeed, applied cryptography and encryption are not the sum of one inventor or invention, they are the product of thousands of years of applied research. The final sections of the chapter explore the growing public demand for encryption and the paradox encryption creates for ransomware victims.

3.1 Applied Cryptography

Kings, organisations and ordinary citizens have long used cryptography and various other forms of coded messages to secure and validate data. The origins of cryptography can be traced back to Leon Battista Alberti in 1466, who is commonly considered to be the "Father of Western Cryptography." Alberti is credited with three significant cryptography advances: "the earliest Western exposition of cryptanalysis, the invention of polyalphabetic substitution, and the invention of enciphered code" (Kahn 1967). Susan Landau, a bridge professor in cybersecurity at the Fletcher School of Law and Diplomacy, defines cryptography as "the science of transforming communications so that only the intended recipient can understand them" (Landau 2004, p. 89). The usage of the terms encryption and cryptography is synonymous within academia, industry and the mainstream media. To avoid confusion and for the purposes of differentiating between classical cryptography and encryption, within this research, encryption is defined as a computer-enabled process that uses an algorithm to convert plain text into a cipher text for the purposes of preventing unauthorised access.

M. Ryan, *Ransomware Revolution: The Rise of a Prodigious Cyber Threat*,
Advances in Information Security 85, https://doi.org/10.1007/978-3-030-66583-8_3

When discussing the influence of encryption on society, Professor Lawrence Lessig of Stanford Law School explains "here is something that will sound very extreme but is at most, I think, a slight exaggeration: encryption technologies are the most important technological breakthroughs in the last one thousand years" (Lessig 1999). According to former FBI Director Louis Freeh, if strong encryption is openly available to the wrong people, "then law enforcement officials will be powerless to stop those people from committing crimes of extraordinary degree" (Freeh 1997). The former director continued by explaining that as far back as 1997, encryption has been an integral part of criminal syndicate planning operations and communications methods.

When analysing the field of applied cryptography, it is essential to acknowledge that cryptography and encryption are diverse fields with fluctuating degrees of complexity. Therefore, it is imperative from the outset to acknowledge that "there are two types of cryptography in this world; cryptography that will stop your little sister from reading your files, and cryptography that will stop major governments from reading your files (Schneier 1996)" – and this research is focused on the latter. Encryption has benefitted from the continued research efforts of Whitfield Diffie and Martin Hellman who originally discovered the concept of public key cryptography in 1975. They were also amongst the first to observe the influence of technology was having on encryption methods identifying that "the development of cheap digital hardware has freed it from the design limitations of mechanical computing and brought the cost of high grade cryptographic devices down to where they can be used in such commercial applications" (Diffie and Hellman 1976). This observation led them to challenge the security of the Data Encryption Standard (DES), arguing that the 56-bit key length was now too short to prevent brute-force attacks. Diffie and Hellman's research would inspire Adi Shamir who has played an instrumental role in numerous encryption advances. Shamir is most known for his works with Ronald Rivest and Leonard Adleman, who were able to practically demonstrate public key cryptography. The trio successfully "showed how a message could easily be encoded, sent to a recipient, and decoded with little chance of it being decoded by a third party who sees it" (Rivest et al. 1983). This research would later become known as RSA encryption, named after its authors. It is one of the most widely used encryption algorithms ever invented, and it is also frequently used to encrypt victim's data and keys in ransomware attacks.

The cybersecurity ecosystem is complex and diverse. There exists a great divide between many government officials, security researchers and academics and the hacking community (which are more accurately defined as either white hat or cybercriminals). Whilst it is difficult to point to a research paper, the influence of Jeff Moss on cybersecurity and applied cryptography should not be understated. Moss originally rose to fame as a hacker who later founded the Defcon and Blackhat cybersecurity conferences. He is a former CSO at Internet Corporation for Assigned Names and Numbers (ICANN) and is currently a Department of Homeland Security Advisory Council (HSAC) member. Moss's continued contribution to cybersecurity has enabled three distinctive groups with conflicting views to come together and share cybersecurity knowledge. It can be argued that these events may have had a

Table 3.1 Schneier's estimated times for a brute-force attack against DES in 1995 (Schneier 1996, p. 153)

	Length of key in bits					
Cost	40	56	64	80	112	128
$100 K	2 seconds	35 hours	1 year	70,000 years	10^{14} years	10^{19} years
S1 M	0.2 seconds	3.5 hours	37 days	7000 years	10^{13} years	10^{18} years
$10 M	0.02 seconds	21 minutes	4 days	700 years	10^{12} years	10^{17} years
S100M	2 milliseconds	2 minutes	9 hours	70 years	10^{11} years	10^{16} years
$1 G	0.2 milliseconds	13 seconds	1 hour	7 years	10^{10} years	10^{15} years
$10 G	0.02 milliseconds	1 second	5.4 minutes	245 days	10^{9} years	10^{14} years
$100 G	2 microseconds	0.1 second	32 seconds	24 days	10^{8} years	10^{13} years
$1 T	0.2 microseconds	0.01 second	3 seconds	2.4 days	107 years	10^{12} years
$10 T	0.02 microseconds	1 millisecond	0.3 second	6 hours	10^{6} years	10^{11} years

positive and negative influence on world's cybersecurity; nonetheless, Moss's efforts have significantly influenced the fields of cybersecurity and applied cryptography. Another well-known hacker who has turned his skills to white hat hacking is Bruce Schneier.

Schneier's early work in the book *Applied Cryptography* is widely considered as the handbook of modern encryption. Since his early works, Schneier has continued to push for improved encryption standards to improve Internet and data security. Within Schneier's *Applied Cryptography,* he demonstrated that the processing requirement for a brute-force attack against a simple symmetric key is relatively easy (Schneier 1996). For instance, if the key is 8 bits long, then there are between 28 and 256 possible keys, and therefore, it will take a maximum of 256 attempts to identify the correct decryption key. If the key is 56 bits long, then there are between 256 and 72,057,594,037,927,936 possible keys. Based on Schneier's assumption that a supercomputer in 1996 could attempt a million keys per second, it would take up to a maximum of 2285 years to identify the correct key. If this is increased to 128 bits long, it would have taken up to 1025 years to decrypt at the time. Since the book *Applied Cryptography* was first published in 1996, encryption has made extraordinary advances; for example, the entire Data Encryption Standard (DES) rainbow table developed by Schneier displayed in Table 3.1 above was superseded just 6 years later by the AES.

3.2 Evolution of Applied Cryptography

Classic cryptography techniques have been used by ordinary citizens for thousands of years. The perpetual requirement for secure communications has led to cryptography remaining in a continuous cycle of improvement and refinement for centuries. In the twentieth century, this process continued to slowly evolve in the shadows until the outbreak of war once again provided cryptography the opportunity to demonstrate its value and power. The success of the German Enigma code raised the profile of cryptography, driving the Allied forces to invest heavily in advanced decryption techniques. The investment eventually paid off with the invention of the electromechanical rotor machine known as the Bombe (Winterbotham 1975). The invention of the Enigma code and ultimate decryption of it provide an enduring reminder why cryptography remains an essential component for secure communications and data storage.

In the period from World War II until the early 1990s, the cryptography domain remained primarily occupied by advanced military powers. The advent of the computer and subsequently the personal computer triggered structural changes in cryptography's application and users. The advent of the personal computer made it possible for advanced military grade encryption techniques to be practised by people with little or no understanding of cryptography. This enabled the broader public to relatively, easily and rapidly apply advanced encryption techniques to their personal communications and stored data. At this stage, it can be argued that there is enough evidence to indicate there is no invention that has changed the field of cryptography more than the advent of the computer. This advancement in conjunction with the rapid evolution of personal computers into smaller, more portable electronic communication devices has led to a dramatic rise in the general public's use of advanced encryption techniques.[1]

Lessig (1999) argues "until recently, there was little non-governmental demand for encryption capabilities. Modern encryption technology – a mathematical process involving the use of formulas (or algorithms) – was traditionally deployed most widely to protect the confidentiality of military and diplomatic communications. However, with the advent of the computer revolution and recent innovations in the science of encryption, a new market for cryptographic products has developed" (Lessig 1999). The rise of the global economy demanded an instantaneous, affordable and secure long-distance communication method. Encryption enabled the Internet through interconnected systems to meet this demand, creating a new primary communication method for business and user communications.

3.3 Public Demand for Encryption

The outbreak of technology and the Internet generated tremendous interest in the field of cryptography. This has led to a treasure trove of academic and publicly available research and software-based tools for applying and attacking almost every variety of encryption. Continuous advances in computational power associated with Moore's law have required users to routinely and substantially increase the

[1] Note: Social media stories such as Wikileaks and Edward Snowden's may have also played a significant role in the broader public's use and perception of advanced encryption techniques. The use of encryption can also be commonly associated with dissidents and citizens under repressive regimes and those involved in forms of illegal activities such as drug trafficking and terrorism.

complexity and size of their encryption access keys (Moore 1975). Inversely, the same advances in computational power have also permitted the same advances for decrypting and hacking encryption keys. Through a process typically referred to as brute-forcing, computer enabled applications such as random key and password generators can be applied against encrypted systems. Fundamentally, this is the most basic example of an automated version of a thief trying to gain unauthorised access to an item protected by a classic three-disc mechanical padlock cipher; the thief, by cycling through the all-number combinations, will inevitably learn the correct access key.[2]

This rapid technology uptake in turn led to the creation of multiple new encryption formats which are typically further subcategorised as symmetric and asymmetric based on the structure of their key systems. Symmetric keys use the same key to encrypt and decrypt, whereas asymmetric algorithms were designed to use a key for encryption (public) and a different key for decryption (private). Typically, the private key is much larger than the public key, and this prevents the encryption key from being used to calculate the decryption key within any reasonable time period (Schneier 1996).

> Standardised encryption algorithms have come and gone as vulnerabilities have been discovered and exploited to make algorithms unusable in the current cryptographic climate. Critical events such as the theoretical cracking of the Data Encryption Standard (DES) and the triple DES led to a global encryption competition that resulted in the broad adoption of the Advanced Encryption Standard (AES) (Fluhrer et al. 2011).

By the mid-1990s, there was an increasing public distrust of DES encryption standard and a growing chorus calling for a new encryption standard to be developed. In response, the National Institute of Standards and Technology (NIST) in 1997 announced a competition for an Advanced Encryptions System (AES) that would run 128-bit blocks of data using 128-, 192- or 256-bit keys. A constraint of the competition was that all encryption designers would have to surrender propriety rights, and the algorithm would become open source (no royalties) (National Institute of Standards and Technology 2001). NIST accepted submissions from around the globe and determined that fifteen had met the criteria. Almost a year later, NIST announced five finalists: MARS, Serpent, Twofish, RC6 and Rijndael. The following year after extensive public consultation and review, NIST announced that Rijndael was to be selected as the winner of the AES. The proposal's name Rijndael is a combination of the two designer's names, Joan Daemen and Vincent Rijmen (Daemen and Rijmen 1998). From a historical perspective, the Rijndael system also applied Kerckhoff's principle: "the system must not be required to be secret; it should be able to fall into the hands of the enemy without causing any inconvenience" (Kerckhoffs 1883). The transparency of the Rijndael system was a significant factor in the widespread acceptance and adoption of the AES.

---

[2] Note: Current generation decryption applications apply extensive data analytics, using the most commonly used passwords first. Statistically, this could drastically reduce the time required to discover the key in comparison to conventional accumulation-type brute-force attacks.

In June 2003, the "U.S. government approved the use of AES 128-Bit for the protection of all documents classified as SECRET and the use of AES 256-Bit for documents classified as TOP SECRET" (National Security Agency 2003). As a further testament to the AES resistance to attack, despite continued advances in technology and computing power, the 2017 version of the Australian government's *Information Security Manual* still recommends the same encryption protocols for the protection of SECRET and TOP SECRET data. A fundamental component of the AES algorithm structure is its almost linear properties that can be scaled to increase the encryption's resistance to brute-force-style attacks with limited effect on encryption/decryption processing times. Additionally, in practical applications where the encryption key is known, it takes relatively the same amount of time and computational power to encrypt a designated block of data as it takes to decrypt the same block of data using a computer of the same computational power.

Over the last 30 years, as computers have become smaller and faster, organisations and individuals have been able to deploy encryption and decryption at will. Today with one touch, a child can encrypt a real-time message on Facebook messenger using 256-bit AES encryption. There is no requirement for the user or the message recipient to understand or even have the faintness comprehension about encryption. This same process is available through a variety of mobile devices and open-source software applications. We live and operate in an environment where criminal organisations can rapidly apply the same types of advanced encryption methods to protect their data as the most advanced military superpowers use to secure their systems. As early as 2004, a time well before open-source mobile devices and applications offered free and virtually instantaneous AES 256-bit encryption, Landau predicted that the "AES algorithm may become the most widely used algorithm of the new century" (Landau 2004, p. 89).

Today, the sheer volume of data being created and encrypted continues to grow exponentially every year. Continuous improvements in technology, software and computational power have facilitated and necessitated stronger encryption algorithms. This has and will ultimately continue to lead to stronger and more advanced methods of encryption, which naturally in turn will lead to larger and larger time periods required to undertake successful brute-force attacks. There is no shortage of proponents who argue for the pros and cons of this endless development cycle.

3.4 No Compromises

An underlying aspect this research aims to highlight is the double-edged sword of advanced encryption techniques. As governments, corporations, academics and cryptography enthusiasts endure to develop and implement new encryption methods for securing and transmitting their data, criminal organisations seek to pilfer these works for own security and financial gain. Technical analysis of ransomware attacks indicates that the underlying source code components that execute the encryption process (key generation and key management) have evolved significantly

Table 3.2 Theoretical maximum times to brute force AES-256 encryption keys (Arora 2012)

AES key size	Duration
56-bit	399 seconds
128-bit	1.02×10^{1S} years
192-bit	1.872×10^{37} years
256-bit	3.31×10^{56} years

in recent times (Kharraz et al. 2015). Understanding the strength of encryption is essential to managing cyber risk and responding to cyberattacks. Those charged with managing cyber risks need to be cognisant and clearly understand what they are going to be up against in the event of a ransomware attack maturing against their organisation.

The decryption challenges ransomware poses can be demonstrated by the time taken to brute force an AES 256-bit symmetric key.[3] This process can potentially be sped up by the application of password dictionaries and rainbow tables.[4] Table 3.2 details the maximum theoretical brute-force times derived by Arora in 2012 using a supercomputer that was capable of maintaining a speed of 10.51 petaflops:

When discussing the standards required for modern encryption, Schneir argues "either build encryption systems to keep everyone secure, or build them to leave everybody vulnerable" (Schneier 2016). Greenberg asserts cryptography was once the realm of academics, intelligence services and a few cipher punk hobbyists who sought to break the monopoly on the science of secrecy. Today, the cipher punks have won; encryption is everywhere. And it's easier to use than ever before (Greenberg 2017). The problem is the more secure our encryption techniques become, the safer our data is, but on the other hand, the more effective ransomware attacks become. The book finishes with a warning: Do not underestimate the skills of the people behind ransomware, they are constantly looking for new ways to exploit weaknesses.[5]

In the first attacks, ransomware typically targeted particular file types, such as JPG, DOC, PDF, XLS, ZIP and a variety of other commonly used file extensions. The next ransomware variants were developed and concealed within fake antivirus programs. Thakkar explains that it is beyond any doubt that cybercriminals are no longer happy with recognition and glory but are focused on their return on investment (ROI). This approach can be demonstrated in the dynamic pricing of ransomware demands, which applies a dynamic geographical pricing model. This assumes that some ransomware attackers decide to adjust the ransom demands based on their perception of the victim's capabilities (Thakkar 2017).

---

[3] Note: This is the absolute theoretical maximum time frame it would take to discover all possible keys.

[4] Note: A rainbow table is a precomputed (calculated) table for reversing cryptographic hash functions.

[5] Note. NotPetya is a clear example of ransomware that was designed to be destructive.

Raju (2017) insists that a defence "in depth strategy is preferable in the long run, as there can never be enough contingency plans with the rate at which technology changes" (Raju 2017). As an example, the use of physically isolated off-site cloud storage platforms will improve recovery options. Although cloud storage and operated services are publicised as one of the safest data storage options, cloud-hosted systems are not immune to the threat posed by ransomware. Cloud-hosted systems may provide distinct advantages for specific attack or recovery scenarios, but they are not a magical bulletproof solution.

3.5 Conclusion

This chapter explored the evolution of applied cryptography throughout history and the ongoing public debate related to advanced encryption techniques. It highlights that modern encryption has benefitted from the combined research efforts of Alberti, Diffie, Hellman, Shamir, Rivest, Adleman, Moss, Schneier and Turing, just to list a few. Whilst it has taken thousands of years for encryption to evolve to this point, the advent of the computers and the Internet have rapidly influenced the speed of encryption transitioning an advanced military technology to an open-source tool. Indeed, despite encryption's progressive strength improvement (complexity and resistance to attack), it has become easier than ever to apply. It is this ease of application that makes encryption so attractive to cybercriminals and ransomware developers.

References

M. Arora, How secure is AES against brute force attacks? *EE Times*. (2012) [Online]. Available online: https://www.eetimes.com/document.asp?doc_id=1279619#. Accessed 10 Mar 2018

J. Daemen, V. Rijmen, *The Design of Rijndael: AES – The Advanced Encryption Standard* (Springer, Berlin, 1998)

W. Diffie, M. Hellman, New directions in cryptography. IEEE Trans. Inf. Theory **22**(6), 644 (1976)

S. Fluhrer, Mantin I., Shamir A., Weaknesses in the key scheduling algorithm of RC4, *Paper presented at the International Workshop on Selected Areas in Cryptography*. 2011

L.J. Freeh, Director of the Federal of Investigations. (9 Sept 1997) [Presentation]

A. Greenberg, How to encrypt all of the things, *WIRED*. (2017). Available online: https://www.wired.com/story/encrypt-all-of-the-things/. Accessed 26 Apr 2019

D. Kahn, *The Codebreakers: The Story of Secret Writing* (MacMillan, New York, 1967)

A. Kerckhoffs, La cryptographie militaire. J. Sci. Milit. **IX**, 5–38 (1883)

A. Kharraz, W. Robertson, D. Balzarotti, L. Bilge, E. Kirda, *Cutting the Gordian Knot: A Look Under the Hood of Ransomware Attacks* (Springer International Publishing, Cham, 2015)

S. Landau, Polynomials in the Nation's service: Using algebra to design the advanced encryption standard. Am. Math. Mon. **111**(2), 89–117 (2004)

L. Lessig, *Code and Other Laws of Cyberspace* (Basic Books, New York, 1999)

G.E. Moore, Tech Digest. IEDM **21**, 11–13 (1975)

National Institute of Standards and Technology, Cryptographic standards and guidelines. (2001). Available online: https://csrc.nist.gov/Projects/Cryptographic-Standards-and-Guidelines/Archived-Crypto-Projects/AES-Development. Accessed 26 Apr 2018

National Security Agency, National Policy on the Use of the Advanced Encryption Standard (AES) to Protect National Security Systems and National Security Information, in *Committee on National Security Systems*, (National Security Agency, Washington D.C., 2003)

K. Raju, Can cloud storage save you from ransomware attacks?, *Info Security Magazine*, 17 July 2017 (2017)

R. Rivest, A. Shamir, L. Adleman, U.S. Patent 4,405,829. Issued September 20, 1983. Available here, *Cryptographic Communications system and method* (United States Patent 1977). Available online: https://patents.google.com/patent/US4405829A/en. Accessed 22 Jan 2020

B. Schneier, *Applied Cryptography: Protocols, Algorithms, and Source Code in C* (Wiley, New York, 1996)

B. Schneier, A 'Key' for encryption, even for good reasons, weakens security, *The New York Times.* 15 July 2016, (2016) [Online]. Available online: https://www.nytimes.com/roomfordebate/2016/02/23/has-encryption-gone-too-far/a-key-for-encryption-even-for-good-reasons-weakens-security. Accessed 26 Apr 2019

D. Thakkar, *Preventing Digital Extortion* (Packt Publishing, Birmingham, 2017)

F. Winterbotham, *The Ultra Secret* (Weidenfeld & Nicolson, London, 1975)

Chapter 4
Ransomware Economics

This chapter explores the underlying economics of cybercrime. The chapter begins by examining how cybercrime syndicates adapt to changing market conditions. This following section of the chapter explores how ransomware demonstrates the ability to monetise both valuable and innocuous data. The chapter then moves forward to discuss how Internet marketplaces have changed the dynamic for criminal activities and why some cybercriminals are shifting their focus to ransomware. It explores why cybercriminals have become more focused on holding files hostage for money than on unleashing stolen data to the black market (Parrish 2018). The final section of the chapter examines what are cryptocurrencies and what impact they have in ransomware attacks.

4.1 Cybercrime Economics

In the book *You'll See This Message When It's Too Late*, Wolff extensively details how ransomware did not begin to emerge as a well-known mode of attack until 2009, which coincides with the same period that Bitcoin and other anonymous modes of online payments became more mainstream (Wolff 2018, p. 71). This creation and broader adoption of anonymous digital payments allowed criminals to shield their profits from being traced or blocked by law enforcement and financial intermediaries. When examining the Gameover Zeus botnet attacks, Wolff makes the following observation:

> That the most enduring legacy of the operation was not the decentralised peer-to-peer bot model…But the economic model of extortion via Cryptolocker, which, just as deliberately removed many layers of money mules and intermediaries from financial transactions. These financial intermediaries who route transactions or fence stolen data were typically weak spots in large-scale cybercrimes that enabled law enforcement intervention (Wolff 2018, p. 69).

M. Ryan, *Ransomware Revolution: The Rise of a Prodigious Cyber Threat*, Advances in Information Security 85, https://doi.org/10.1007/978-3-030-66583-8_4

The online and open-source nature of cybercrime suggests that researchers have for decades had the opportunity to study the operations and organisation of cybercrime actors. Singer and Friedman (2014) argue that this is possible "by infiltrating digital black-markets, where criminals trade the necessary components of their schemes. Forum sellers post offers for spam, credit-card numbers, malware, and even usability tools" (Singer and Friedman 2014, p. 90). However, few researchers have analysed why organised cybercrime syndicates may have switched operating tactics and the influence emerging technologies were having on their criminal business operations.

This convergence of technology and financial systems is a fundamental requirement in modern ransomware attacks motivated by money. Garrick Hileman (2017) from the London School of Economics insists that "cryptocurrencies are the result of a combination of multiple achievements in various disciplines that include, but are not limited to computer science (P2P networking), cryptography (cryptographic hash functions, digital signatures) and economics (game theory)" (Hileman and Rauchs 2017). These technological advances represent some of the key underlying components that make ransomware such an attractive option for organised cybercrime syndicates. Kesari et al. (2017) argue that "like ordinary businesses, financially motivated cybercrimes are an activity of scale, not a jackpot activity such as robbing a bank. Criminals need to optimize their processes, make sales, and critically, they rely on many different intermediaries for everything from marketing, to web hosting, to delivery of products" (Kesari et al. 2017).

The Harvard Business Review proposes that the broader community should think of cybercriminals as business enterprises because, fundamentally, most just want to make a profit. They explain "rather than thinking of a clandestine hacker working out of a basement, you will be better served to picture a sophisticated, professional operation working out of an office tower" (Gardiner 2017). The criminal industry is a well-oiled, and thriving machine. Their operations often require partnerships, specialisations and logistical supply chains. To succeed long term, these criminal enterprises will need to understand when to share information and when to exploit it. Ultimately, many of these criminal enterprises are reliant on conducting business with other criminal enterprises. Navigating that terrain successfully requires a certain nous because as the old Roman proverb states, there is simply no honour among thieves.

4.2 Economics of Ransomware

Ransomware emerged on a global scale at a point in time when stolen financial-related data was abundant and declining in value. Not dissimilar to other criminal markets such as narcotics, cybercrime is not immune to supply and demand market forces. The global adoption of new digital technologies drove businesses to move at unprecedented rates from pen and paper files to digital documents. The alluring promise of increased speed, efficiency and profits inspired organisations to rapidly adopt new Internet-based technologies to conduct business communications and transactions. This rapid adoption of technology left many organisations unprepared and unprotected to the type of cyberthreats they would encounter. This increased level of insecurity left many enterprises exposed and large volumes of data were stolen, which led to an abundance of stolen data becoming available to buy. This led to the price falling year-on-year, leaving cybercriminals to search for new opportunities to make profits – enter ransomware.

4.2.1 Monetising Data

The emergence of any emerging commodity touting lucrative returns is a certain catalyst to prompt government regulators to intervene in an effort to restrain or prevent those who control its flow. This occurred more than a century ago with oil, and today, similar concerns are being raised about the new oil – big data (The Economist 2017). Whilst Internet technology giants such as Google, Amazon, Microsoft and Facebook quietly began devising how they could collect, analyse and monetise data at unprecedented rates, organised crime syndicates were also contemplating new strategies to improve the scale and profitability of their lucrative online operations.

Organised cybercrime syndicates rapidly explore emerging technological advances for their moneymaking potential. Despite operating only in the virtual shadows, cyberattackers display technical skills in conjunction with business acumen and a level of self-awareness. They not only understand sound security practices such as "the value of the data must remain less than the cost to break the algorithm protecting the data. The time it takes to break the algorithm must also be greater than the value of the data (Schneier 1996)" but reverse-engineer these best practices to their own advantage. This agile economic approach is reinforced by Yale business Professor Rodrigo Canales who argues that organised crime syndicates often adopt and exhibit sophisticated and innovative economic models that are consistent with leading multinational corporations (TED 2013).

From the outset of cybercrime activities such as fraud, identity and credential thefts, cybercriminals knew that proceeds of crime-associated financial transactions initiated from Russia, Eastern Europe and other developing countries had a low probability of success. To counter these traditional defences, institutional regulations and law enforcement, organised criminal syndicates recruited money mules within the countries where the stolen data or credit-card holder was located. Money mules are commonly enlisted as intermediaries for criminal organisations to transport and process fraudulently gained money to criminals in a different location. Many are even unaware of their involvement with criminal syndicates. This approach is adopted because locally based mules generally draw less attention from law enforcement, lowering the risk of the operations being discovered. In cybercrimes, mules have frequently been used to exploit stolen credit cards by making transactions in local banks before on-sending the money to the parent criminal organisation abroad (Kshetri 2010, p. 1070). This type of economic model is lucrative but is a labour-intensive process requiring a complex web of mules.

In the period since 2011, the volume of stolen data available on the Internet has increased. The primary reason for this is governments, corporations and individuals are simply producing and collecting more data than ever before. Quid pro quo, there is more data available to steal, so naturally, there is more stolen data available on the black market. Examination of black markets during this period also indicates that the value of stolen data was rapidly decreasing, thanks to the overwhelming supply. In the period of 2011 to 2016, the price of a stolen credit card on the black market dropped from a high of $25 to a low of $6 (USD) (Verizon 2016). An alternative study undertaken by Trustwave in 2017 found that the mean value of basic PII was $0.03 cents, social security details were $0.50, access credentials $0.95 and credit cards were $5.40 (USD) (Trustwave 2017).

In an interesting segue, the researchers from Trustwave also found that enterprise security and risk professionals exponentially overestimate the market value of personal data for sale on the black market. Typically, the data value is less than 5 percent of the value estimated by senior IT managers. For a payment card records, “senior IT managers over-estimate the value by 60 times the actual criminal values of data for sale on the black market” (Trustwave 2017). The estimated data value for a single banking record was even less accurate with the average estimate being 2000 times higher than the actual value. This trend in declining stolen data values was starting to negatively impact criminal’s return on investment strategy, at the same time improvements in technology were making it easier for corporations and law enforcement to analyse millions of smaller transactions.

> So not only were financially motivated criminals limited in what types of data they could monetise, but they were making smaller and smaller profits from selling their stolen data. Ransomware solved both the problems – enabling criminals to extract value from all sorts of seemingly worthless data stored on people’s devices for much higher fees than they had ever been able to charge on the black market (Wolff 2018, p. 70).

Ransomware attacks fundamentally changed the economic model for many organised cybercrime syndicates. Instead of wasting finite resources discretely breaking into systems, searching for valuable data and then exfiltrating data, much less sell it through any online forum; ransomware simplified their criminal operations. This shift eliminated money mules and several key bottlenecks that defenders and law enforcement might try to disrupt. By storing the encrypted data directly on the victim’s own computers and selling it back directly to them for cryptocurrencies, criminals were largely able to escape powerful third parties with strong incentives to intervene (Wolff 2018, pp. 77–78).

When we consider cybercrime as a process, arguably, the greatest risk comes at the point of sale, not the point of theft. In the early twenty-first century (and still even today), stealing valuable data was not the most difficult task in the process in the cybercrime cycle. For example, once a series of credit-card information has been stolen, the thief cannot simply place an advert on craig’s list advertising the sale. Financial institutions have become acutely aware of the value of threat intelligence, including the emergence of stolen data related to their entity in online marketplaces. This is further complicated by financial institutions and law enforcement agencies who now actively share threat intelligence information they discover.

This creates a complex scenario where the thief needs to monetise the data before the financial institution becomes aware of the theft and cancels the cards. Additionally, the attacker will most likely need to validate a portion of the data with the purchaser before finally receiving payment. The final payment may come in the form of an anonymous cryptocurrencies, but this still needs to be converted to cash or some other form of currency beyond the purview of law enforcement before the profits can be realised. The challenge of selling stolen data is not insurmountable, but it does present a series of risks which threaten the entire criminal operation.

It has been argued that financial intermediaries who route transactions or sell stolen data have typically been considered weak spots in large-scale cybercrimes that enabled law enforcement intervention (Wolff 2018, p. 71). Traditional fiat currencies and financial institutions have provided choke points where law enforcement has successfully been able to leverage, exploit and disrupt cybercrime operations. In 2014, the US Department of Justice (DOJ) even launched "Operation Choke Point" which specifically "targeted certain merchant categories that were identified as being at high-risk for money laundering and consumer exploitation (scams, payday lenders, etc.)" (Silver-Greenberg 2014).

The existing financial system remains an Achilles' heel for many organised crime syndicates. Frequently, their exploits have been exposed, and profits seized long after all the hard work has already been done. Law enforcement and security researchers have frequently been able to infiltrate black markets to identify large caches of stolen data, often reporting the stolen data to the owners before it could be used or sold. Even after a successful sale, cybercriminals may endure considerable problems trying to launder their profits to conceal their activities and identity. Money laundering continues to be a complex problem, and technology is arguably increasing the complexity of this challenge. Whilst technology opens new avenues to launder money, it has also removed many of the exhaustive labour process associated with forensic investigations. Modern financial systems and forensic tools provide extensive data analytics, meaning what may have taken investigators months in years gone past can now be completed in seconds.[1]

There are also significant profit degradations through the laundering process. Cybercriminals commonly sell their services and stolen data for anonymous gift cards. These anonymous gift cards are then generally on sold to another party at a significantly reduced rate. This cycle may occur multiple times before the attacker feels comfortable to exchange the anonymous cards or product for fiat currencies. Whilst this process can add protection through transaction complexity, it may result in an attacker only receiving a fraction of the initial stolen data's sale price. Another research study by Kharraz et al. (2015) also found that:

---

[1] Note: Data analytics tools may provide significant improvements in analysing known data sets. An example of this would be analysing internal company financial records for theft or fraudulent transactions. However, currently, these data analytics tools cannot be (or fully) applied to analyse transactions and exchanges between all known fiat currencies and cryptocurrencies. This indicates that attackers retain an advantage when utilising automated (scripted) transfers across multiple currencies and cryptocurrencies to launder illicit funds.

> A large fraction of ransomware samples (88.22 percent) used prepaid online payment systems such as MoneyPak, Paysafecard, and Ukash cards, since they provide limited possibilities to trace the money. These services are not tied to any banking authority and the owner of the money is anonymous. The ransomware business model takes advantage of these systems since there are no records of the vouchers to trace cybercriminals (Kharraz et al. 2015).

In this process detailed in Fig. 4.1, the point of conversion to fiat currencies represents the highest risk point. While other cyberattackers or law enforcement may disrupt or prevent the sale of the stolen data, the point of conversion provides an optimum time to identify the attacker. This identification process may rapidly lead to the profits being seized or the attacker being apprehended by law enforcement. Whilst somewhat effective, this economic model creates numerous laundering-related problems for cyberattackers.

Researchers from Princeton University and Google developed multiple methodologies to track ransomware payments from end to end. The researchers successfully identified payments in excess of $16 million (USD) linked to the "Locky" and "Cerber" ransomware attacks; however, the researchers ultimately concluded that they were unable to determine all linked payments derived from the "Locky" attacks. The analysis also identified cryptocurrency exchanges as a possible choke point for law enforcement; however, the researchers cited the difficulty in trying to identify the true identities of the account holders even at the physical exchanges (Huang et al. 2018).

This is the point where cryptocurrencies become an attractive option. Not only are cryptocurrencies difficult to trace, but they are also attractive because they are highly portable. An attacker can easily route millions of dollars' worth of cryptocurrencies through digital platforms such as emails, messenger and text message. It also provides attackers a simple method to potentially transport millions of dollars from the proceeds of organised crime across international borders by carrying a USB drive. A recent Cambridge University research study observed that:

> Initially, a cryptocurrency exists in a vacuum; a closed system that has no connections to other systems (e.g., other cryptocurrency systems, traditional finance, the real economy). In order to participate, users need to start mining in order to earn the cryptocurrency, which can only be used for transacting with users of the same system as there is no way to spend or sell them. To counter this, exchanges are established that let users trade cryptocurrency for other cryptocurrency and/or national currencies (Hileman and Rauchs 2017).

Black markets and cryptocurrency laundering platforms that exchange cryptocurrencies with cybercriminals can request commission fees of up to 50 percent.[2] The exchange or black market then resells the virtual coins or vouchers at a further discount to its end users. This process is simplified above in Fig. 4.2. Though an attacker loses a percentage of the ransom earned, the attacker is still left with a quite handsome amount. The transactions performed in such markets are usually untraceable (Thakkar 2014).

---

[2] Note: This practice is not dissimilar to stolen property in the physical world. Due to the attention the property (coins) will draw from law enforcement, this increases the risk associated with the exchange. For accepting this burden, the exchange demands increased returns.

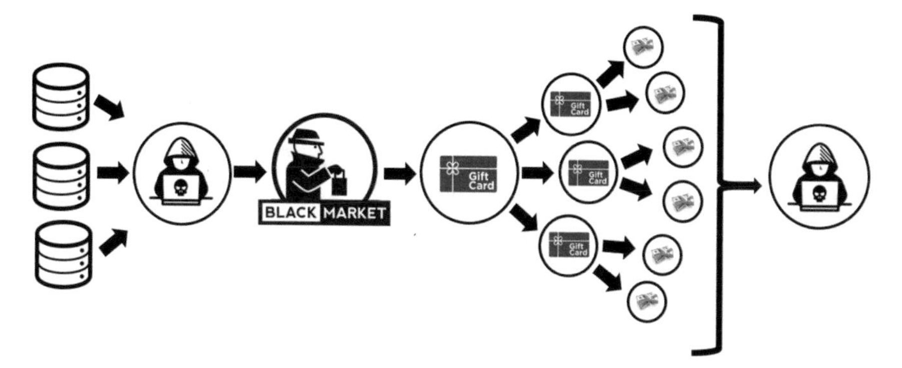

Fig. 4.1 Example of stolen data sale process

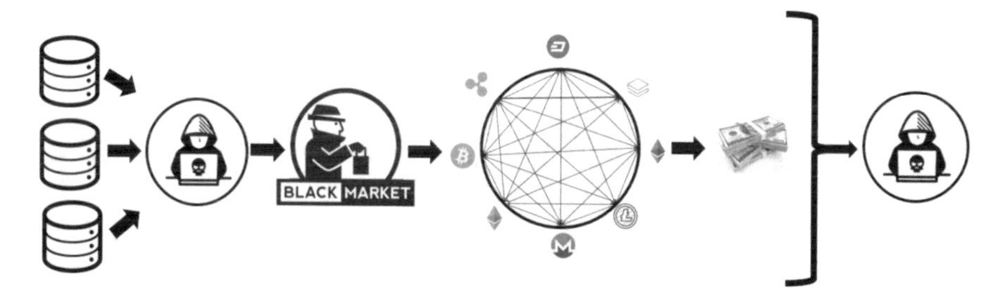

Fig. 4.2 Example of stolen data sale process using cryptocurrencies

> Pseudo-anonymity and irreversibility of Bitcoin transaction protocol have made Bitcoin a dexterous utility among cybercriminals. Unlike genuine users, who seek to transact securely and efficiently; cybercrooks exploit these characteristics to commit immutable and presumably untraceable monetary fraud (Conti et al. 2018).

This process is further enhanced using the modern ransomware model detailed below in Fig. 4.3.

Recent research from Cambridge University estimates that there are up to 11.5 million wallets (accounts) that are currently active. Almost 40 percent of wallet providers offer multi-cryptocurrency exchange services, with another 30 percent of providers planning or in the process of rolling out this feature by 2020. The research also identified that all fiat-to-cryptocurrency exchanges perform anti-money laundering (AML) checks of wallet holders, with the preferred method being a process of internal checks (Hileman and Rauchs 2017).

4.2.2 Request for Payment

The Ponemon Institute's analysis of ransomware payments indicates that Bitcoin is the most requested ransomware payment currency (Ponemon Institute 2017). Other major cryptocurrencies demanded include Bitcoin Cash, Ethereum and Monero. As

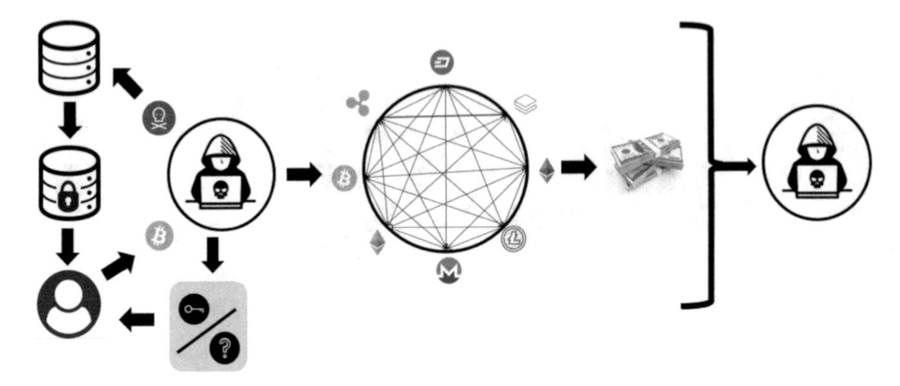

Fig. 4.3 Example of simplified ransomware process using cryptocurrencies

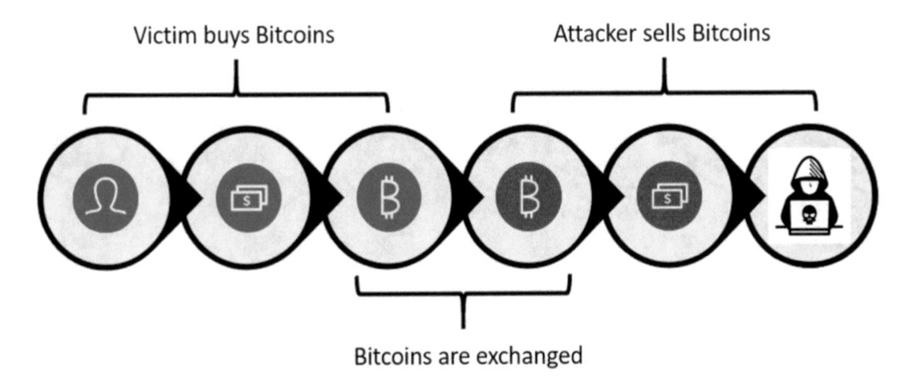

Fig. 4.4 Example of ransomware payment cycle

the most prominent cryptocurrency used in ransomware attacks, Bitcoin continues to draw the greatest attention from mainstream media and law enforcement. However, security researchers are increasingly observing cybercriminals transitioning from Bitcoin to Monero due to its enhanced privacy features over Bitcoin. Major privacy features of Monero include Ring Confidential Transactions (RingCT), stealth addresses, decoy coins and cryptographic design that prevents unauthorised users from seeing the account balance within a given wallet. Figure 4.4 details the most commonly observed ransomware payment cycle:

Outside cryptocurrencies, other anonymous payment methods including prepaid cards such as Paysafecard, Ukash, MoneyPak or cashU have been requested in ransomware attacks. Ransom demands are usually requested in virtual currencies to inhibit or prevent law enforcements ability to trace the attackers via financial records. Further analysis of ransom demands indicates there is no exact pattern to determine the ransom amount, with recent ransom demands varying from $100 to $2000 (USD) (Thakkar 2014, p. 125). However, there is some anecdotal evidence that suggests that targeted ransomware attacks typically demand higher ransoms.

> Many key cybersecurity decisions are based on a risk/reward basis. For some companies, it is financially prudent to pay for the cost of an attack rather than invest unknown dollars in preventing one. What undermines the risk/reward ratio is the assumption of understanding the price of data. How much is a personal social security number or credit card actually worth to an illegal buyer? How much money will an attacker ask for to release a piece of ransomware? This information is important when determining a risk-to-reward cybersecurity strategy to protect data vulnerabilities. You might assume a specific piece of information is worth 75 cents when on the Dark Web it could sell for $1.25 or 25 cents. Either way, you are operating under a giant assumption around value instead of validating it in a real Dark Web setting (Cohen 2017, p.55).

4.2.3 Payment Predicaments

Many security experts urge organisations to prepare and plan defences against ransomware attacks. Despite these continued warnings, the evidence indicates many people and corporations continue to fall victim to ransomware attacks. Moreover, there is substantial evidence indicating that many people and corporations continue to pay ransom demands. The decision to pay a ransom demand can be challenging, and there are numerous driving forces behind why people pay ransom demands. These forces include fear of exposure, fear of uncertainty, shame, convenience and the fear of losing valuable or sentimental data.

Another contributing factor is probability of data recovery. While there are no certainties when dealing with organised cybercriminals, recent research data indicates that ransomware victims have a 55–70 percent probability of recovering their data after they paid the ransom demands (Trend Micro 2018; Ponemon Institute 2017). Whilst far from certain, it does highlight that the profitability of ransomware is somewhat predicated on trust and gaining repeat business. The ability to infect more victims is primarily reliant on the technical prowess of the attacker's malware, and the ability to generate further revenue will be shaped by factors such as the victims' perception, related experience and vulnerability.

Public perception and related experiences may play a large part in a victims' decision to pay the ransom demand.[3] Ransomware attackers are reliant on the infamy of their attacks. At the same time, if making money is the objective of the attack, then developing a brand-like reputation for granting immediate access once payment is made is influential to future victims' decision-making processes. When the next user becomes a victim, their immediate actions are to call their service provider or their IT friend and to trawl the Internet for answers. Not dissimilar to legitimate businesses, user feedback and reviews may provide a strong influence in a victim's decision to pay the ransom demand.

When users or corporations become victims of ransomware, many face an ethical question about paying the ransom. Many victims ponder where their ransomware

[3] Note: A family member, friend or co-worker may have discussed a previous ransomware experience. This may have a significant influence on the victims perceived situation and its potential outcomes.

payment is ultimately going to end up. There is no simple answer to this question, and where ransom payments ultimately arrive is diverse and unclear. There is a logical rationale that ransom payments represent a return on investment for cybercriminals and their operations, with a percentage being taken as income and the remainder being reinvested in their next criminal operation. In this scenario, victims contemplate whether paying the ransom perpetuates the cycle of organised crime, inevitably leading to further cyberattacks in the future. In other scenarios, ransomware attacks have been attributed to rogue states as a method of generating revenue to circumvent international sanctions (Bossert 2017). This may lead some victims to question whether their payment is supporting terrorism or state-sponsored violence.

For large corporations, the decision to pay or not pay should be a decision that is already made prior to any ransomware or cyberattack.[4] Corporations that have a mature cyber defence strategy and capability should know their potential exposure and risk thresholds for common cyberattack scenarios. The decision to pay should be a business decision derived from exposure, cost and potential damage to the brand. This is not to say the decision process will be easy or straightforward. Ransomware attacks such as the one incurred by Maersk demonstrate ransomwares has the potential to rapidly spread through complex enterprises wreaking havoc beyond standard business continuity plans (Greenberg 2018). For governments, banks and security providers, ransomware attacks can be embarrassing and have devastating impacts on their business operations and reputations.

For others, the decision to pay the ransom may be relatively easy. The ransom amount requested may be of a small monetary value, and the compromised data or system may be of a high monetary or personal value to the victim or corporation. From a practical perspective, individual users and small-to-medium enterprises may have no other option to recover their data other than paying the ransom. A common scenario is that the victim may have stored photos or personal data that is only stored on the infected device. The victim may even have a boot disk to restore the system; however, the restoration process would restore the system but result in the loss of the personal data. With law enforcement unable to offer any practical assistance, and the victim having limited resources and time against them, victims have no choice but to pay the ransom if they wish to recover their data.

Understanding why victims pay ransoms can range from being a simple to complex process. Whilst some victims may refuse to pay on principle, these decisions become increasingly complex for large enterprises. The time and cost of recovering may be significant, and the access or loss of data in infected systems may have catastrophic impacts on a business's ability to operate. For others, paying the ransom may present an ethical dilemma they are unable to accept or overcome. It's easy for security experts to postulate that corporations should be prepared for ransomware attacks; in reality, preparedness in large corporations can be a complex and resource-sapping process.

---

[4] Note: Corporations should also understand their legal and regulatory constraints. In many Western countries, paying a ransom may not be permitted due to Know Your Customer (KYC) and anti-money laundering (AML) restrictions.

4.3 Black Markets

Internet-based black and darknet markets quickly emerged when cybercriminals and hackers became aware that a lot of money could be made by trading hacking techniques and stolen data. The rapid growth of the Internet provided a universal platform for like-minded Internet users to interact and to trade goods and services. Initially, black markets were considered to provide a low-risk platform for many hackers and criminals because law enforcement was ill-equipped to track transactions and monitor their communications and remains relatively slow to adapt to the demands of governing cyberspace (Ablon et al. 2014b). Even when law enforcement agencies have had success in closing darknet markets, buyers and sellers have almost instantaneously moved on to the next one (Hanuka 2015).

Moving forward, Ablon et al. argue that "black markets will require better encryption, vetting, and operational security due to the dynamic give-and-take between black-market actors, law enforcement and security vendors. Participants will employ innovative methods and tools to help obfuscate, encrypt, or make a transaction quicker, easier to use, and harder to find" (Ablon et al. 2014b). Despite having limited success in bringing down black markets, there are concerns from pockets of the law enforcement community that high-profile takedowns such as the Silk Road may be driving criminal entities even further below ground. With every high-profile takedown, organised cybercrime syndicates feel that law enforcement is closing in, further driving them to continuously review and improve their own security measures and processes.

4.3.1 Black and Darknet Markets

Defining or describing the entire darknet market is a difficult proposition because it is so vast and disjointed, has too many users and is constantly evolving. The composition of users is also constantly evolving, with many experts estimating that "in the mid-2000s, approximately 80 percent of black-market participants were freelance (the rest being part of criminal organisations or groups)" (Ablon et al. 2014b). However, this has sharply declined with many experts now estimating the freelance market is closer to 20 percent today. The evolution of darknet markets has also created tiers (layers) that mirror legal marketplaces. Access to low-tier marketplaces is generally considered open, whilst access to high-level tiers is to be considered heavily restricted. Access to these high-level tiers often requires personal connections and reputation to establish trust before access will be granted. These high-level areas are virtually policed and controlled by internal hierarchies, with system administrators restricting access to invited users only.

Ablon et al. further detail that that open-source channels and chats have now been replaced by "participants hosting their own servers, sharing email accounts where the content is exchanged by saving draft messages, and using off-the-record messaging, the encryption scheme GNU Privacy Guard (GPG), private Twitter accounts, and anonymizing networks such as Tor, Invisible Internet Project (I2P), and Freenet" (Ablon et al. 2014a, p. 7). Threat actors over the last 5 years have increasingly become sophisticated and innovative in their communication methods. These evolving business practices have in turn increased their level of obscurity, making their activities more hidden and anonymous.

The darknet is full of marketplaces selling everything from legal items to illegal drugs, stolen personal information, fake passports, software exploits and even university degrees. The darknet is broadly considered to be an "anonymous cyberspace where at least half of its visitors are selling illegal information, stolen data, codes, drugs, pornography, and weapons" (Cohen 2017). Figure 4.5 below illustrates a sample RaaS for purchase online.

A 2016 study into digital underground markets by researchers at RAND Corporation discovered "there's even a kind of brand-name hierarchy, Russian hackers have a reputation for quality. Some Vietnamese groups focus on e-commerce. Chinese hackers are known for targeting intellectual property, and Americans tend to specialize in financial crime" (Irving 2016). Some cybersecurity experts have even hypothesised that the black market can be more profitable than the illegal drug trade: Links to end users are more direct, and because worldwide distribution is accomplished electronically, the requirements are negligible. This is because the majority of players, goods and services are "online-based and can be accessed, harnessed, or controlled remotely, instantaneously. Shipping digital goods may only require an email or download, or a username and password to a locked site. This enables greater profitability" (Ablon et al. 2014a, p. 11).

These black markets provide a platform for cybercriminals to trade exploits and techniques that facilitate the further development of new ransomware attacks. The goods and services available on darknet markets are limitless, enabling cybercrimi-

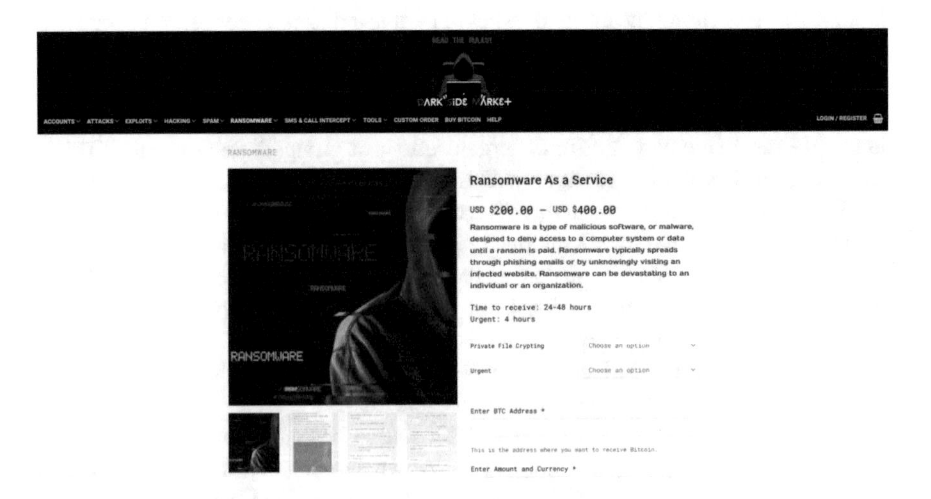

Fig. 4.5 Example of RaaS in marketplace

nals to rapidly procure previously unknown security vulnerabilities. Even Internet users with limited or no hacking skills can purchase pre-made attacks (exploit kits) and rent organised cybercriminals to undertake cyberattacks such as ransomware attacks. Darknet markets provide Internet users direct access to nefarious actors and their goods and services.

4.3.2 Darknet Forums

The darknet's composition in many ways mirrors the mainstream regular Internet; there are forums and marketplaces.[5] Darknet forums provide a virtual meeting point for users to discuss topics within a community of other like-minded users. There typically are no geographical, age or sex restrictions, with forums often comprised of diverse users from around the globe. For non-darknet users, Reddit is a synonymous mainstream Internet platform that enables similar community discussions (threads) and sub-Reddits. Both platforms are structured to have admin (moderators) and users, with admins having additional privileges which are akin to a bar having bar staff and patrons. To ensure user safety and privacy, forum administrators often require user login and implement various other security mechanisms to prevent unauthorised access to the site (Robertson et al. 2016).

Forums are an essential component of the darknet, and forum access is categorised as open or restricted to members only. Whilst some forums will block users from listing or discussing topics such as child pornography, weapon sales and human trafficking, the darknet has traditionally been considered the Internet's wild west (a lawless virtual society). One key difference to the mainstream Internet is the lack of search engines (Google, Yahoo, etc.); however, this is replaced by linked list. A linked list is essentially a list of darknet sites with embedded or website addresses commonly posted on a forum. Forums also allow users to discuss and rate other users, such as sellers. If a user is going to shell out their hard-earned Bitcoin for molly (pure ecstasy), then it is essential that they trust the seller. Before its closure, the Silk Road allowed users to report and rate sellers like they were reviewing a purchase on eBay or Amazon.

Darknet forums provide a platform for cybercriminals to discuss their exploits – both past and planned. This anonymous setting allows like-minded users to exchange information. This exchange of information can occur through anonymous channels with no oversight from law enforcement. This provides an open forum for criminals to discuss hacking techniques and the ability to draw on expertise from around the world. This enables developers of ransomware attacks to orchestrate attacks that exceed their own networks' technical prowess. Ross Ulbricht, the founder of the Silk Road, was reliant on forums and other darknet users to configure and maintain

[5] Note: The Internet is usually accessed through search engines. The darknet and Dark Web are the same distinction. The darknet is the network of computers that you can't usually see/search, and the Dark Web is the system that allows you to interact with them.

the security of the Silk Road. Based on the value of trades, the Silk road was valued at 2.4 billion in 2013 when Bitcoin was $120 per coin – Bitcoin was worth over $20,000 (AUD) in 2017.

4.4 Cryptocurrencies

Throughout history, currencies have essentially provided a receipt for a commodity – redeemable in most cases for physical gold (European Central Bank 2012). Today, most physical currencies are considered fiat currencies, which are typically provided by governments, and are controlled by some form of central authority. These currencies have been controlled and regulated by these central authorities, and the value of the currency can be derived from the trust placed in the currency by its users (Kien-Meng Ly 2014). In recent times, pockets of the community have become sceptical about the control measures and roles played by these central authorities. Because of this growing distrust in conjunction with a "phenomenon triggered by technological developments and by the increased use of the internet, communities have sought to create their own currencies" (European Central Bank 2012).

The Internet provides a ubiquitous platform to exchange information, goods, services and currency with varying degrees of anonymity and transparency. The increasing shift towards Internet-based financial transactions is the by-product of globalisation, consumer dissatisfaction, speed of change and user convenience. The Internet has accentuated the role many governments and major financial institutions play, triggering many to question the underlying basis for these roles. These considerations have led academics and the global Internet community to begin researching and developing alternative options outside the existing global financial system.

The quest to develop alternative banking platforms and solutions is not confined to technology hubs like Silicon Valley or small pockets of merchants and consumers on the Internet. In the largest study of its kind, EY's Banking Relevance Index Report, which surveyed 55,000 consumers across 32 markets, found an "increasing interest in banking alternatives" (EY 2016, p. 3). The survey's findings detail that "Banks have historically played an important part in people's lives. While consumer demand for financial services will continue, it is unclear to what extent traditional banks will provide these services in the future" (EY 2016). Proponents of Bitcoin's system view it as a workaround for their lack of trust in the existing payment infrastructure, which is dependent on governments, imperfect central banks or payment intermediaries (Angel and McCabe 2015).

Many of the alternatives developed and currently being developed have sought to establish a new financial system that removes trusted third parties from commerce on the Internet, which in the existing global financial system are almost exclusively comprised of government-backed financial institutions (Nakamoto 2008). Another driving force for creating an alternative option is the delay time it currently takes for international financial transactions to be processed. In the age of cryptocurrency, Michael Casey argues that it is quicker to board a flight in London and fly to

Singapore than it is to wire transfer money through the international banking system between these locations (Casey 2018). This example highlights the frustrations many multinational organisations, international traders and merchants share over the existing international banking system.[6]

As the number and types of alternative financial exchange currencies and platforms continue to grow, blockchain- and cryptocurrency-based (also known as virtual or digital currency) financial systems have emerged as their respective market leaders for opposing reasons. Despite the rise of both systems, there remains a high degree of consumer and regulatory confusion about the different currencies and their associated platforms. The most common consumer perplexity surrounds the differences between Bitcoin and blockchain. A simple explanation is that Bitcoin can be considered an application that uses blockchain as its underlying operating system. Whilst there are fundamental similarities, there are also differences between the two (Lucas 2018). The basic characteristics of Bitcoin are as follows: an open source, unregulated, fully digital currency that requires complex encryption to secure, process, and validate its distributed ledger across a public network, which in-turn enables every user to view every transaction that has ever occurred. As a result, Bitcoin payment transactions are generally considered to be anonymous and require no personal identifiable information (PII) (Bryans 2014).

On the other hand, blockchains are characterised as private, access controlled and regulated through a central repository, which used a transactional database that is distributed across a trusted P2P network. The system is permission based so that users may only see transactions they are permitted to see. Blockchains may also be used to exchange or track items of value such as information, goods, intellectual property and currency. Whilst blockchain proponents proclaim a myriad of advantages, which may be indeed true, the system further ingrains and enhances the government and financial institution oversight, the fundamental component many of its creators sought to eradicate. It is also the fundamental reason cryptocurrencies like Bitcoin will continue to prosper for criminal activities.

4.4.1 Coercion through Simplification

The adoption of technologies that deploy advanced encryption techniques has become so simplistic that ransomware perpetrators can coerce their victims too use the technologies as well. There is an underlying paradox in the ingenuity of ransomware perpetrators to coerce their victims not only to pay their ransom demands but to purchase cryptocurrencies to make the payments anonymous.[7] Within three decades, the point of entry to adopt advanced encryption technologies has reduced

[6] Note: Cryptocurrencies are also generally not bound by regulations such as KYC and AML, which may cause delays in conventional finacial systems.

[7] Note: To increase the probability of payment, some ransomware attackers have created even detailed instructions and inbuilt help desk-style features that assist victims purchase and make payment using cryptocurrencies.

in level from mathematician to average computer user. Concurrently, request for ransom payments has evolved from mailing cheques to Panamanian postboxes to electronically purchasing and transferring anonymous cryptocurrencies to anyone anywhere in the world instantly.

The use of cryptocurrencies as the preferred payment method has reduced the risk to the perpetrators whilst simultaneously providing geographic agility. Ransomware attackers have the ability to remotely launch attacks, whilst being free to move locations, without being constrained by the need to wait for the ransom payment to arrive. Ransomware payments can be made across international borders, without delay and without oversight from central banks and law enforcement. Profits from the attacks can then be withdrawn or exchanged from anywhere at any time.

4.4.2 Bitcoin

Cryptocurrencies have emerged as the prominent digital exchange platform. Presently, there is no shortage of mainstream cryptocurrency platforms which include Ripple (XRP), Litecoin (LTC), Monero (XMR), Ethereum (ETC) and Bitcoin (BTC). There are also numerous Initial Coin Offerings (ICO) in development and scheduled in the coming years, with a Chinese local government even launching its own $1.6 billion USD blockchain (Aslam 2018). Determining which currency is the most used and has the largest market value is subject to market volatility. However, Bitcoin is presently the most renowned cryptocurrency with global Internet users, and it is also the most commonly requested payment platform for ransomware attacks.

Bitcoin was founded by Satoshi Nakamoto, which many experts believe is a pseudonym to conceal the real identity of the creator or creators. The system was announced in 2008, with the system's structure first appearing in the White Paper *Bitcoin: A Peer-to-Peer Electronic Cash System.* The paper defines a new cryptocurrency-based financial system that is described as:

> A purely peer-to-peer version of electronic cash would allow online payments to be sent directly from one party to another without going through a financial institution. Digital signatures provide part of the solution, but the main benefits are lost if a trusted third party is still required to prevent double spending. We propose a solution to the double-spending problem using a peer-to-peer network. The network timestamps transactions by hashing them into an ongoing chain of hash-based proof-of-work, forming a record that cannot be changed without redoing the proof-of-work. The longest chain not only serves as proof of the sequence of events witnessed, but proof that it came from the largest pool of CPU power. As long as a majority of CPU power is controlled by nodes that are not cooperating to attack the network, they'll generate the longest chain and outpace attackers. The network itself requires minimal structure. Messages are broadcast on a best effort basis, and nodes can leave and re-join the network at will, accepting the longest proof-of-work chain as proof of what happened while they were gone (Nakamoto 2008).

Bitcoin and other digital currencies seek to create a financial system that is not based on conventional trust but instead on mathematics, cryptography and distributed networks. Through a complex mathematical process, a public transaction led-

Table 4.1 Cryptocurrency transactions per day in 2017 (Hileman and Rauchs 2017)

Period	Bitcoin	Ethereum	DASH	Monero	Litecoin
Q1 2016	201,595	20,242	1582	579	4453
Q2 2016	221,018	40,895	1184	435	5520
Q3 2016	219,624	45,109	1549	1045	3432
Q4 2016	261,710	42,908	1238	1598	3455
Jan–Feb 2017	286,419	47,792	1800	2611	3244

ger is created and updated approximately every 10 minutes. The ledger is stored publicly and is dispersed through millions of nodes around the world. Table 4.1 below highlights the increasing popularity of cryptocurrencies as a medium to exchange money since 2016. The ledger contains the data about every single transaction, cross referencing and verifying each transaction. However, a key difference with Bitcoin is the fields of information contained within the ledger. Whilst the transactions ledgers are public information, the account owner's information is not. There is no central repository containing the account user's personal identifiable information. There is no account name, age, sex, geographical marker or any other identifying details. Accounts are created solely online, require no identification and are anonymous.[8] To further increase the security of accounts, many Bitcoins accounts are only used once. This means that one person or organisation may potentially own thousands of different Bitcoin accounts, with no digital method to link or trace these accounts to one owner.

An exception to this is the transcribing of these accounts into physical or into other data ledgers. Whilst many hackers, criminals and cyberterrorists employ extensive security measures when operating online, many continue to utilise physical ledgers for storing Bitcoin and other e-wallets account information. These physical and data-based ledgers have subsequently been discovered during investigations and through the execution of search warrants by law enforcement agencies (McMillen 2013). In the United States, once ledgers are recovered by law enforcement, and the case has been successfully prosecuted, the perpetrators' accounts are frequently sold back to the market (Ciolli 2017).

With an increased level of anonymity, it should come as no surprise that cybercriminals, terrorists and Transnational Organised Crime (TOC) syndicates have begun using Bitcoin and other cryptocurrencies to fund their illegal activities. Analysis of the Bitcoin system indicates that all transactions are viewable through the public ledger where Bitcoins can then be traced to their extraction point. The monitoring and tracking process can be further complicated through the usage of

[8] Note: A user accessing the Internet from a public connection may be able to improve their level of online anonymity; however, they may inadvertently increase their probability of detection and identification through third-party systems such as CCTV. A user accessing the Internet from a private Internet connection that is routing through a VPN in combination with software applications such as TOR can generally expect their Internet connection and online activities to be anonymous. The obvious exception is a user's system and networks whose hardware or software integrity has already been compromised (i.e., under surveillance by the state, a state agency or another actor).

"Bit laundry" services. Bit laundry services seek to unlink and mix Bitcoins across multiple accounts, significantly reducing the ability for Bitcoins to be tracked. In 2013, Moser performed an analysis of the anonymity and transaction graph of three Bitcoin mix services. He found that all the three of the Bitcoin mix (laundry) services had a distinct transaction graph pattern, but some of them were more successful than others at obfuscating the Bitcoin's true origins (Moser 2013).

4.4.3 Ransomware and Bitcoin Tracing

It has been suggested that Bitcoins used in the payment of ransoms could be digitally marked; however, this suggestion has been widely disputed in the global Bitcoin market. The fundamental reason that many users elect to undertake transactions using virtual currencies such as Bitcoin is based on the virtual currency's unfettered movement and high degree of anonymity. There are several academic studies that have successfully traced Bitcoins; these studies have focused primarily on tracing Bitcoins used in legal marketplaces or from users of the defunct *Silk Road*. Whilst these studies demonstrate the ability to trace some coins, they have been unable to link these transactions with a physically identifiable user (Goldfeder et al. 2017). As a result, many of the research methodologies used to date have limited or no application with tracing ransomware payments.

In an international study of 1359 types of ransomware between 2006 and 2014, researchers from Symantec, Northeastern University, Lastline Labs and Institut Eurecom analysed the payments of the Cryptolocker[9] ransomware attacks:

> The analysis of the transactions shows that cybercriminals started to adopt evasive techniques (e.g., using new addresses for each infection to keep the balances low) in order to better conceal the criminal activity of the Bitcoin accounts. Our analysis also confirms that the Bitcoin addresses used for malicious intents share similar transaction records (e.g., short activity period, small Bitcoin amounts, small number of transactions). However, determining malicious addresses in the Bitcoin network based on the transaction history is significantly difficult, in particular when cybercriminals use multiple independent addresses with small number of Bitcoins (Kharraz et al. 2015, p. 2).

The researcher also observed that a significant fraction of these Bitcoin addresses (68.93 percent) were active for a maximum of 10 days. Whilst discarding the account doesn't erase the record of the transaction from the public ledger, it does reduce the risk of attribution.

> Bitcoin provides some unique technical and privacy advantages for miscreants behind ransomware attacks. Bitcoin transactions are cryptographically signed messages that embody a fund transfer from one public key to another and only the corresponding private key can be used to authorize the fund transfer. Furthermore, Bitcoin keys are not explicitly tied to

---

[9] Note: The Cryptolocker attacks infected over 500,000 machines between 2013 and 2014, using primitive spam messages to spread the virus and RSA encryption to lock the user's files before demanding payment. The virus was ultimately brought down by a white hat campaign Tovar.

real users, although all transactions are public. Consequently, ransomware owners can protect their anonymity and avoid revealing any information that might be used for tracing them (Kharraz et al. 2015, p. 14).

Applying the same method in which conventional criminals use burner mobile phones, once the account has been used and the funds extracted, there is no value in retaining or maintaining the account, and it is effectively discarded. Once a victim pays the ransom, the collected Bitcoins are commonly transferred through a series of temporary intermediate accounts. In turn, the coins in these accounts are split into numerous small amounts and transferred to a series of new accounts. This process is often repeated multiple times before all the coins are recombined into a new account. This process increases the complexity of tracing the money and increased the resources required to undertake the tracing process.

An investigation by Google researchers analysed 301,588 different strains of ransomware across 34 known families. Using this data, the researchers were then able to examine Bitcoin transactions against the blockchain ledger with known ransomware wallets. The research team excluded results that were not determined to be of a high degree of confidence, with the remaining results able to trace $25 million (USD) since 2014. Analysis by Brewster (2017) uncovered that "95 percent of ransomware profits were cashed out at the Russian exchange BTC-E. That chimes with the indication that the biggest ransomware types are the products of the biggest organised cybercrime syndicates working out of Russia" (Brewster 2017).

4.5 Changing Market Conditions

Throughout 2016–2017, ransomware remained the top cyberthreat on all major global cyberthreat indexes; however, in early 2018, ransomware was replaced by cryptojacking.[10] The profitability of ransomware was creating a crowded market, and by late 2017, some organised crime syndicates began making a shift towards cryptojacking attacks. Whilst this shift caused the number of ransomware families to decrease, the number of ransomware variants actually increased by 46 percent (Symantec 2018). This change also indicates that syndicates were focused on short-term returns and less focused on innovations. The shift towards cryptojacking can be primarily attributed to soaring global cryptocurrency prices. In late 2017, a sharp rise in cryptocurrency prices directly altered the profitability of cryptojacking. This triggered cybercrime syndicates to rapidly shift focus from ransomware attacks to cryptojacking attacks.

It can be argued that cryptojacking represents a return to more traditional forms of cybercrime – its objective is to discretely steal resources from its victims over a prolonged period of trying without drawing attention to the attack. The priority for

[10] Note: The use of the term cryptojacking refers to the process of using malware to illegally steal CPU revolutions from victims' devices to mine cryptocurrencies. (See Fuscaldo 2018).

this type of attack is to remain undetected, making minimal changes to the victims' system to avoid drawing any suspicion of attack or unauthorised access. The discrete nature of the attack allows the attacker to remain in the victims' system for a prolonged period of time, increasing the potential returns from their illegal crypto mining activities. The passive nature of the attack also reduces the probability of law enforcement interdiction, lowering the risk to the attacker. Stealing CPU cycles from a large financial institution is unlikely to draw the same level of ire from law enforcement versus encrypting the institution's data and disrupting their core ability to operate.

Unfortunately, the rise of cryptojacking may have limited or no medium- to long-term effect on ransomware attacks. Whilst cryptocurrency prices remain relatively high, cryptojacking will be an attractive low-risk option for cybercriminals. As a result, the rise of cryptojacking may reduce the number of ransomware attacks by reducing the number of ransomware attacks being developed and deployed, but this change will have no impact on preventing, defending or responding to ransomware attacks. It may even splinter parts of the cybercrime market, leaving ransomware to upper echelon malware developers who continue to make strong profits from ransomware attacks.

For organisations, the rise of alternative cyberthreats such as cryptojacking does little to reduce the prodigious threat posed by ransomware. Whilst cryptojacking poses a financial threat to organisational resources, the impacts from cryptojacking are not comparative with ransomware attacks. The removal of cryptojacking malware may be as simple as running antivirus software or reinstalling an Internet browser. The rise of cryptojacking to the top of multiple cyberthreat indexes is misleading and does not accurately reflect the impacts from ransomware attacks (IBM 2019; Symantec 2018). The shift also fails to recognise that ransomware attacks are not always driven by profit. For nation states, ransomware may have multiple applications including coercion, value degradation, disruption and warfare. The potential anonymous application of ransomware ensures that nation states will continue to pursue the further development of ransomware attack for their future warfare purposes.

4.6 Conclusion

This chapter explored the shifting cybercrime economics landscape. The rapid adoption of new digital technologies created new opportunities to steal date, which led to an oversupply of stolen data in black markets. This led to security professionals, financial institutions and government agencies to develop cyberthreat intelligence capabilities, which in turn has increased the risks associated with selling stolen data. Cybercriminals were quick to adapt to changing market conditions, with many shifting from a data theft to a data access denial model (ransomware). This shift in business has proven lucrative for many, with the advent of cryptocurrencies aiding the profitability of this transition.

References

L. Ablon, M. Libicki, A. Golay, Projections and predictions for the black market, in *Markets for Cybercrime Tools and Stolen Data Book Subtitle: Hackers' Bazaar*, (RAND Corporation, Santa Monica, 2014a)

L. Ablon, M. Libicki, A. Golay, Characteristics of the black market, in *Markets for Cybercrime Tools and Stolen Data Book Subtitle: Hackers' Bazaar*, (RAND Corporation, Santa Monica, 2014b)

J. Angel, D. McCabe, The Ethics of Payments: Paper, Plastic, or Bitcoin? J. Bus. Ethics **132**, 603–611 (2015)

N. Aslam, Bitcoin: The harder the fall, the higher the rise: $35K by Q4, *Forbes*. (2018) [Online]. Available online: https://www.forbes.com/sites/naeemaslam/2018/04/10/bitcoin-the-harder-the-fall-the-higher-the-rise-35k-by-q4/#18e1126d5410. Accessed 10 Apr 2018

T. Bossert, Press briefing on the attribution of the WannaCry malware attack to North Korea, 19 Dec 2017

T. Brewster, Google warns ransomware boom scored crooks $2 million a month, *Forbes*. (25 July 2019) 2017 [Online]. Available online: https://www.forbes.com/sites/thomas-brewster/2017/07/25/google-ransomware-multi-million-dollar-business-with-locky-and-cerber/#758974576caf. Accessed 17 Jan 2019

D. Bryans, Bitcoin and money laundering: Mining for an effective solution. Indiana Law J. **89**, 440–472 (2014)

M. Casey, How Bitcoin and Blockchain are changing the world. (5 Mar 2018) [Video]. Available online: https://vimeopro.com/user12220083/osmosis-work-samples-for-unsw/video/151457603. Accessed 24 Sep 2019

J. Ciolli, People are making a fortune buying government-seized bitcoins, *Business Insider Australia*. (2017). Available online: https://www.businessinsider.com.au/bitcoin-price-government-auction-winners-2017-5?r=US&IR=T. Accessed 28 May 2018

A. Cohen, Cyber (in)security decision-making dynamics when moving out of your comfort zone. Cyber Def. Rev. (Army Cyber Institute) **2**(1) (Winter), 45–60 (2017)

M. Conti, A. Gangwal, S. Ru, On the economic significance of ransomware campaigns: A bitcoin transactions perspective. Comput. Secur. **79**, 162–189 (2018)

European Central Bank, *Virtual Currency Schemes* (European Central Bank, Germany, 2012)

EY, The relevance challenge: What retail banks must do to remain in the game, *EY*. (2016)

D. Fuscaldo, Crypto mining malware grew 4,000% this year, *Forbes*. (2018). Available online: https://www.forbes.com/sites/donnafuscaldo/2018/12/28/crypto-mining-malware-grew-4000-this-year/#7efbb86a224c. Accessed 13 Sept 2019

M. Gardiner, To guard against cybercrime, follow the money, *Harvard Business Review*. (2017). Available online: https://hbr.org/2017/05/to-guard-against-cybercrime-follow-the-money. Accessed 16 June 2018

S. Goldfeder, H. Kalodner, D. Reisman†, A. Narayanan, When the cookie meets the blockchain: Privacy risks of web payments via cryptocurrencies, *Princeton University*. (2017)

A. Greenberg, The untold story of NotPetya, the most devastating cyber attack in history, *WIRED*. (2018). Available online: https://www.wired.com/story/notpetya-cyberattack-ukraine-russia-code-crashed-the-world/. Accessed 23 Jan 2019

T. Hanuka, The untold story of silk road, part 2: The fall. (2015). Available online: https://www.wired.com/2015/05/silk-road-2/. Accessed 27 May 2018

G. Hileman, M. Rauchs, *Global Cryptocurrency Benchmarking Study* (University of Cambridge, 2017). Cambridge Centre for Alternative Finance, Cambridge, United Kingdom

D.Y. Huang, M.M. Aliapoulios, V.G. Li, L. Invernizzi, E. Bursztein, K. McRoberts, J. Levin, K. Levchenko, A.C. Snoeren, D. McCoy, *Tracking ransomware end-to-end*, IEEE symposium on security and privacy, 23 May 2018 [Video]. Available online: https://www.youtube.com/watch?v=H5bPmzsVLF8. Accessed 14 Mar 2019

IBM, IBM X-Force threat intelligence index 2019. (2019)

D. Irving, The digital underworld: What you need to know, *RAND Review*. (2016), 21 May 2018

A. Kesari, C. Hoofnagle, D. McCoy, Deterring cybercrime: Focus on intermediaries. Berkeley Technol. Law J. **32**(3), 1131 (2017)

A. Kharraz, W. Robertson, D. Balzarotti, L. Bilge, E. Kirda, *Cutting the Gordian Knot: A Look Under the Hood of Ransomware Attacks* (Springer International Publishing, Cham, 2015)

M. Kien-Meng Ly, Coining Bitcoin's "Legal-Bits": Examining the regulatory framework for bitcoin and virtual currencies. Harvard J. Law Technol. **27**(2) (Spring), 588–608 (2014)

N. Kshetri, Diffusion and effects of cyber-crime in developing economies. *Third World Q.***31**(7), 1057–1079 (2010)

M. Lucas, The difference between Bitcoin and Blockchain for business. Available online: https://www.ibm.com/blogs/blockchain/2017/05/the-difference-between-bitcoin-and-blockchain-for-business/. Accessed 12 Mar 2018

R. McMillen, Who owns the worlds biggest bitcoin wallet? The FBI, *WIRED*. (2013). Available online: https://www.wired.com/2013/12/fbi-wallet/. Accessed 14 Mar 2018

M. Moser, Anonymity of bitcoin transactions: An analysis of mixing services, *Monster Bitcoin Conference* 17 Sept 2013, 2013

S. Nakamoto, Bitcoin: A peer-to-peer electronic cash system, (2008). Available online: www.bitcoin.org. Accessed 28 May 2017

K. Parrish, Hackers are now favoring ransomware over personal data theft, *Digital Trends*. (2018). Available online: https://www.digitaltrends.com/computing/report-suggests-hackers-leaning-towards-ransomware/. Accessed 2 Jan 2019

Ponemon Institute, *The Rise of Ransomware*. (2017)

J. Robertson, A. Diab, E. Marin, E. Nunes, V. Paliath, J. Shakarian, P. Shakarian, Darknet mining and game theory for enhanced cyber threat intelligence. *Cyber Def. Rev.***1**(2 (Fall)), 95–122 (2016)

B. Schneier, *Applied Cryptography: Protocols, Algorithms, and Source Code in C* (Wiley, New York, 1996)

J. Silver-Greenberg, Justice department inquiry takes aim at banks' business with payday lenders, *The New York Times*. 26 Jan 2014 (2014) [Online]. Available online: https://dealbook.nytimes.com/2014/01/26/justice-dept-inquiry-takes-aim-at-banksbusiness-with-payday-lenders/. Accessed 22 Jan 2019

P. Singer, A. Friedman, *Cybersecurity and Cyberwar: What Everyone Needs to Know* (Oxford University Press, New York, 2014)

Symantec, *Cryptojacking Skyrockets to the Top of the Attacker Toolkit, Signaling Massive Threat to Cyber and Personal Security*, 21 Mar 2018

TED, *The Deadly Genius of Drug Cartels* (TED Salon, New York, 2013). Available online: https://www.ted.com/talks/rodrigo_canales_the_deadly_genius_of_drug_cartels/discussion?ga_source=embed&ga_medium=embed&ga_campaign=embedT. Accessed 26 Jan 2019

S. Thakkar, Ransomware – Exploring the electronic form of extortion. *Int. J. Sci. Res. Dev.***2**(10), 123–126 (2014)

The Economist, The world's most valuable resource is no longer oil, but data, *The Economist*. (2017). Available online: https://www.economist.com/leaders/2017/05/06/the-worlds-most-valuable-resource-is-no-longer-oil-but-data. Accessed 27 Jan 2018

Trend Micro, What Happens When Victims Pay Ransomware Attackers? 10 Dec 2018. Available online: https://blog.trendmicro.com/what-happens-when-victims-pay-ransomware-attackers/. Accessed 1 Jan 2018

Trustwave, *The Value of Data: A Cheap Commodity or Priceless Asset?* (Trustwave, United Kingdom, 2017)

Verizon, *Data Breach Investigations Report* (2016). Available online: www.verizonenterprise.com/verizon-insights-lab/dbir/2016. Accessed 9 Jan 2018

J. Wolff, *You'll See This Message When It Is Too Late: The Legal and Economic Aftermath of Cybersecurity Breaches* (The MIT Press, Cambridge, 2018)

Chapter 5
Ransomware Case Studies

This chapter examines four major ransomware cases, with the first major ransomware attack in 2013 being used as a template for developing an influx of attacks since 2016. The individual case studies were chosen based on their global impact on organisations and high-profile media reports surrounding the attacks.[1] The case study analysis process analysed the attack methodology and the outcome of each attack to determine similarities and evolutionary changes between each subsequent attack. The analysis also sought to detail the method and sophistication level of each attack, the encryption process and request for payment. These components provide the foundation for further understanding the rising threat posed by ransomware in later chapters.

5.1 Gameover Zeus

In September 2013, the Gameover ZeuS botnet developers re-engineered a version of the ZeuS Trojan that was originally built using a decentralised peer-to-peer (P2P) botnet infrastructure (Jarvis 2013). The botnets' developers released the updated version which included another piece of malware called Cryptolocker. The Cryptolocker malware was rapidly spread using the hundreds of thousands of computers that were already part of the Gameover ZeuS bot infrastructure. Throughout the botnet's life cycle, the ZeuS developers made a number of updates to the underlying source code over the years to improve its functionality and resilience against takedown attempts. Originally, cybercriminals were deploying the Gameover ZeuS botnet to obtain valuable "data such as personal information, passwords, credit card

[1] Note: Four case studies were deemed to be an appropriate number to accurately demonstrate the evolution of major ransomware attacks profiles over a six-year period.

M. Ryan, *Ransomware Revolution: The Rise of a Prodigious Cyber Threat*,
Advances in Information Security 85, https://doi.org/10.1007/978-3-030-66583-8_5

numbers, customer data, confidential commercial information or any other data that related to banking" (Alazab 2015).

The revised version of the Gameover ZeuS bot was re-engineered to steal money in a completely different manner than the original credential stealing malware. Instead of capturing financial account credentials, the Gameover Zeus bot used Cryptolocker to discretely encrypt the hard drives of computers it infected. Once the drives were encrypted, the malware would then demand that the victims make ransom payments if they ever wanted to access their data again.

This shift in attack methodology was fruitful, with analysis of Bitcoin's logs in correlation with the designated Bitcoin accounts, indicating that in one 2-month period between October 15 and December 18 in 2013, roughly $27 million (USD) was deposited into the Bitcoin accounts (Wolff 2018, p. 64). The success of these extortion-based attacks would certainly influence cybercriminals who were already trying to conceive their next cyberattack in a period of heighted government surveillance and declining credit-card values.

5.1.1 Attack Methodology

It is estimated that between mid-2013 and mid-2014, Cryptolocker infected more than 260,000 computers worldwide. Victims included the Swansea Massachusetts Police Department (which paid $750 (USD) to recover its investigative files). Another victim was a Pittsburgh insurance company who refused to pay and eventually spent $70,000 (USD) on recovering its data (Lucas 2015, pp. 99–100). Analysis of attack design and empirical observations associated with its distribution method indicate that Cryptolocker was specifically targeting English-speaking users (such as those in the United States). These figures are supported by the computer manufacturer Dell, which estimated that the cybercriminals responsible made $30 million (USD) in 100 days (Jarvis 2013). Further analysis of the attack vectors indicates the attacks were non-discriminate and almost fully automated, thus allowing the attackers to target tens of thousands of machines simultaneously (Lucas 2015).

It is the first known Internet-based ransomware attack to successfully implement all three of the properties defined by Gazet (Gazet 2010). The user's files were "encrypted using AES with a random key which is then encrypted utilising a 2048-bits RSA public key. The corresponding private key, needed to decrypt the AES key, can be obtained by paying the ransom" (Palisse et al. 2016). Additionally, the Gameover Zeus botnet contained a clever inbuilt Domain Generation Algorithm (DGA) fail-safe mechanism to prevent it being easily shut down. The DGA was configured to produce 1000 domains per week. Molloy (2014) explains that the "DGA produced long, nonsensical strings at one of six top-level domains: .com, .net, .org, .biz, .info, and .ru that could then be registered and used to send commands to regain control of the botnet" (Molloy 2014). This design feature enabled the botnet operators to retain control of their botnet even if the peer-to-peer (P2P) infrastructure was compromised. This functionality significantly increased the complexity of law enforcement's attempts to shut down the botnet.

In October 2013, researchers from Secureworks observed "Cryptolocker was being distributed by the P2P malware Gameover Zeus in a typical pay-per-installation arrangement. In this scenario, Gameover Zeus was distributed by the Cutwail spam botnet using lures consistent with previous malware distribution campaigns" (Jarvis 2013). The malware is multistaged, first downloading and executing Gameover Zeus, before then downloading and installing additional types of malware such as Cryptolocker. Figure 5.1 below illustrates an example of a phishing email designed to target Australians. The message appears to be an official email sent from the Australian government; however, instead, the email conceals multiple links that, once activated, begin discretely downloading malware onto the user's device.

Cryptolocker typically infected user devices through spear phishing campaigns, watering hole attacks and drive-by downloads (Thakkar 2014). Once on the user's device, Cryptolocker begins discretely initiating a command-and-control action sequence, including taking actions to remove its virtual footprints and retain a persistent presence in the event of a reboot. Until this action is successfully completed, the malware is designed to remain hidden to avoid alerting the user or the system

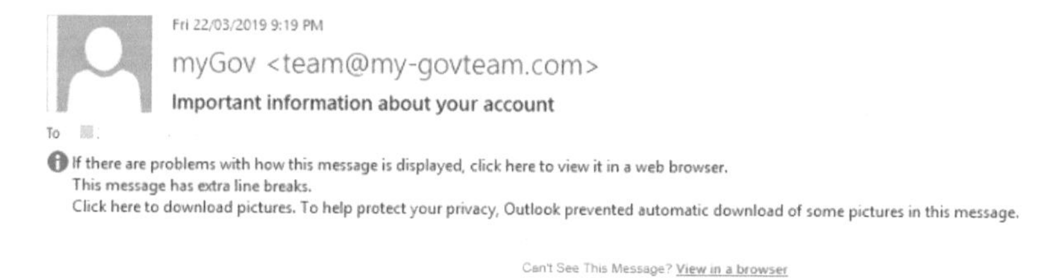

Fig. 5.1 Spam email containing the malware downloader (Australian Tax Office 2020)

tools of its presence. The encryption process is initiated, with Cryptolocker creating an "autorun" registry key. An additional aspect of the attack is that "instead of using a custom cryptographic implementation like many other malware families, Cryptolocker uses strong third-party certified cryptography offered by Microsoft's Crypto Application Programming Interface (API)" (Jarvis 2013).

Through the sound implementation and design practices, the malware developers of Cryptolocker were able to create a robust program that is difficult to circumvent once installed. The difficulty is because "each file is encrypted with a unique AES key, which in turn is encrypted with the RSA public key received from the C2 server. The encrypted key, a small amount of metadata, and the encrypted file contents are then written back to disk, replacing the original file" (Jarvis 2013). To decrypt the system files, a private RSA private key is required, which is exclusively known by the attacker.

5.1.2 *Request for Payment*

Analysis of media reports, academic journals, corporate reports and governments reports indicates that all known WannaCry ransom requests for payment were sought through the cryptocurrency Bitcoin. Once the user's device was successfully infected with the Cryptolocker ransomware, request for payment demands were generally made for $100 (USD) with some victims reporting fluctuations as high as $750 (USD) (Wolff 2018, p. 63). Victims were typically given 72 hours to pay using either anonymous prepaid cash vouchers such as cashU, Ukash and Green Dot MoneyPak or the cryptocurrency Bitcoin (Thakkar 2014). An example demand for payment process is detailed in Fig. 5.2.

Despite offering a discount for ransom payment made through Bitcoin, many victims struggled to figure out how to make Bitcoin payments, so the developers set up a customer service website with step-by-step instructions explaining to the victim how to create a Bitcoin account and make payment. This additional effort by the attackers demonstrates a clear desire to further reduce risk whilst maximising profits. The Cryptolocker developers understood that Bitcoin-based payments could not easily be tied to specific financial accounts and that US law enforcement had limited or no ability to control or monitor these accounts, thus alleviating the risk of the accounts being identified, frozen or seized (Wolff 2018, p. 64).

5.1.3 *Resolution and Attribution*

The Gameover Zeus botnet was eventually traced to authors in Russia and Eastern Europe. Cyberattacks from these regions frequently target victims in Western Europe, Australia and the United States. Part of the reason for this is due to limited communication channels and low levels of cooperation between law enforcement in these regions. These shortfalls can be linked to broader political complications and

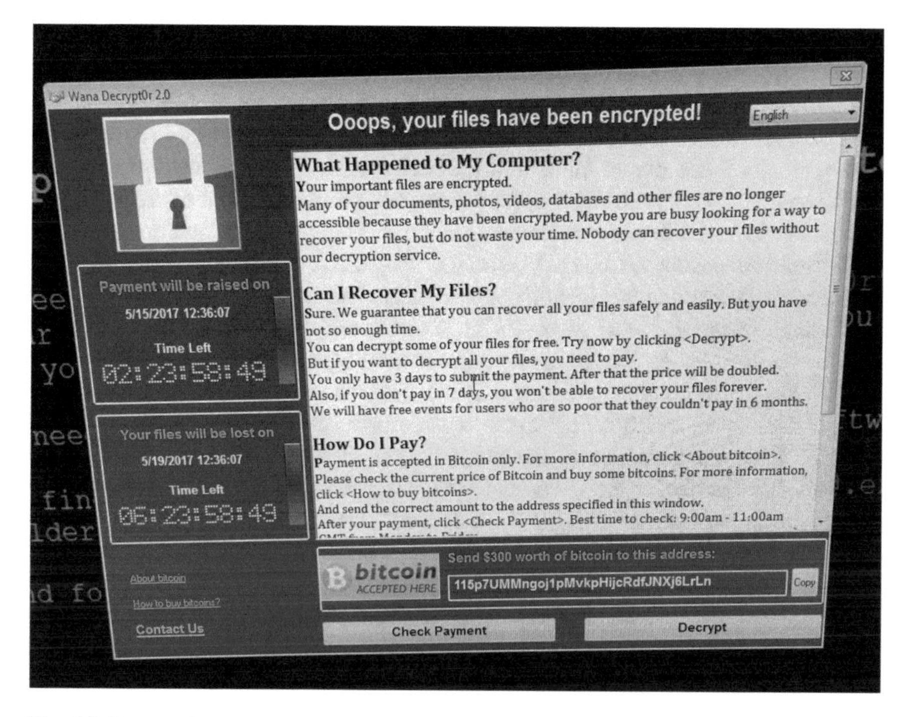

Fig. 5.2 Request for payment

are well known by organised cybercrime syndicates. As a result, this lowers their perceived risk of apprehension and prosecution. Additional reasons why users in these countries are frequently targeted by ransomware attackers include their perceived ability to pay ransom demands and the high number of user devices (larger attack surface) in operation.

Despite this perceived impunity, by late 2013, a coordinated effort was underway between US law enforcement agencies with their Canadian, French, German and Ukrainian law enforcement counterparts. The multinational task force launched operation Tovar to investigate and dismantle the botnet. Forensic analysis of the botnet led to law enforcement agencies to allege "a group of Russian and Ukrainian hackers, led by Evgeniy Mikhailovich Bogachev of Russia" (Molloy 2014; Garber 2014). Despite multiple law enforcement and cybersecurity researcher sources citing Bogachev as the ring leader of the Gameover Zeus botnet, Bogachev has never been arrested by Russian officials.

5.1.4 Results

Whilst official data varies between reports, sources indicate there were up to "3.6 million computers infected by ZeuS in the USA alone during the period of 2009 and 2010" (Etaher et al. 2015). Security researchers say the botnet has led to financial losses of up to $100 million (USD) in the US alone, with one victim reportedly losing $6.9 million (Garber 2014). Analysis of the ransomware attacks indicates that the success of the attacks is reliant on emerging technologies such as encryption, cryptocurrencies and Internet anonymity. Whilst it can be argued Internet anonymity eventually faded due to the work of security researchers and law enforcement, the alleged attackers were never apprehended and prosecuted, and the ransom payments and financial losses were never recovered. These types of outcomes support organised cybercrime syndicates' belief that they are immune or at a very low risk of apprehension and prosecution from foreign law enforcement.

Federal Bureau of Investigation (FBI) executive assistant Robert Anderson stated that the "Gameover Zeus is the most sophisticated botnet the FBI and our allies have ever attempted to disrupt" (Anderson 2014). The malware developers made sound design decisions, which complicated law enforcements efforts to mitigate this threat. They were also able to develop a robust distribution system utilising the Cutwail and Gameover Zeus botnets. The success of Gameover Zeus botnet to launch the Cryptolocker attack provided a potential catalyst for a new wave of financial cybercrime focused on extortion rather than credit-card fraud or identity theft. Migrating payments to cryptocurrencies was a shift that freed criminals from many of the legal constraints and law enforcement oversight capabilities that are used to monitor and prevent suspect fiat currency transactions within financial institutions (Wolff 2018, p. 60).

5.2 WannaCry

In 2017, ransomware remained a significant cyberthreat to governments, corporations and individuals. As the frequency and complexity of ransomware attacks continued to increase, many organisations begun bolstering their cyber defences and resilience efforts to prevent and recover from these types of cyberattacks. Despite a myriad of warnings and a growing threat of ransomware, many organisations failed to take the necessary actions to secure and safeguard their systems. Simultaneously, the cycle of cybercriminality continued to evolve, which saw ransomware attacks increasingly become more sophisticated, more challenging to detect, and harder to recover from. In May of 2017, these shortcomings would come to light again with the release of WannaCry.

The WannaCry (also known as WannaCrypt, WannaCrypt0r and WCry) ransomware attack began to take hold on Friday, May 12, 2017. The US Department of Homeland Security was one of the first government agencies to report the ransomware outbreak, releasing a security alert stating, "a widespread ransomware campaign is affecting various organizations with reports of tens of thousands of infections in over 150 countries, including the United States, United Kingdom, Spain, Russia, Taiwan, France, and Japan" (Department of Homeland Security 2017).

Initially, news organisations and cybersecurity experts linked the attack to an email-based malicious spam campaign. However, after an extensive technical review of data associated with the ransomware attacks, Malwarebytes Labs research indicated that the WannaCry attacks had not been distributed via an email malicious spam campaign. Instead, their research indicated the worm had utilised an Internet-based application that searches for vulnerable public facing Server Message Block (SMB) ports. The attacker(s) had then re-engineered the leaked NSA exploit “EternalBlue” to secure access to the networks, where subsequently they were able to execute the “DoublePulsar” exploit to establish a foothold in the user’s system allowing for the installation and execution of the WannaCry ransomware (McNeil 2017). Security service providers and researchers from around the world raced to identify the attack vector and methods. Their results quickly confirmed Malwarebytes Labs research, tying the ransomware outbreak to the leaked NSA exploits EternalBlue and DoublePulsar (Newman 2018).

5.2.1 Attack Methodology

At the outset of the attack, many cybersecurity experts were quick to express that the attack represented a significant increase in technical sophistication in comparison to previous global ransomware attacks. The WannaCry attack spread rapidly, and within 3 days, it had infected more than 200,000 systems across 150 countries (Greenberg 2018b). Of those infected, the United Kingdom’s health system was amongst the worst affected. The attack had infected or caused disruption to 34 percent of Trusts (service providers) and had infected another 595 local general practitioner’s IT systems (National Audit Office 2018, p. 8). Hospitals throughout the UK reported that the cyberattack was causing large disruptions to their ability to provide medical services, and in some affected areas, the broader public were advised to only seek medical care for emergencies.

Early reports about the event indicated the “National Health Service (NHS) experienced hobbled computer and phone systems, system failures, and widespread confusion after hospital computers started showing a message demanding that a ransom be paid in Bitcoin (Newman 2017)” Later, it would come to light that the attack had affected pathology test results, telecommunications systems, X-ray imaging systems and patient administration systems (National Health Service 2017).

Within days, the malware had infected over 230,000 computer systems in 150 countries, ravaging computers at hospitals in the United Kingdom, rail systems in Germany, universities in China and even auto plants in Japan. In Australia, it was reported that 55 speed cameras had been infected after a technician used an infected USB whilst undertaking maintenance on the system (McLean 2017). Despite being relatively quiet since the initial outbreak, in early 2018, *Forbes* reported that Boeing’s systems were currently under attack by the ransomware, a full year after

Microsoft released the highest priority security patch for vulnerable systems (Matthew 2018).

Analysis of the market sectors, victims, geographical locations, coding, languages and an array of technical research undertaken by multiple independent cybersecurity researchers and laboratories indicates that WannaCry was not created or directed at a specific target. It is possible that there may have been a specific target embedded within those affected; however, an overwhelming body of evidence indicates that the attack was designed to produce a simple widespread, low-cost, indiscriminate cyberattack with the intention to make financial gains. From a technical perspective, the "WannaCry malware consists of two distinct components, one that provides ransomware functionality and a component used for propagation, which contains functionality to enable SMB exploitation capabilities" (Berry et al. 2017).

FireEye's analysis of WannaCry encryption indicates a multistage and multi-algorithm encryption process. Initially, WannaCry generates a new AES key for each individual user file that it intends to encrypt on the target system. WannaCry then "generates a new RSA key pair (only one for each victim), to encrypt all the victim's AES keys on the victim's system, and it subsequently encrypts the unique victim RSA 2048-bit asymmetric private key with the shipped RSA public key" (Berry et al. 2017). To complete the process, WannaCry sends the cipher text (victim's unique encrypted RSA private key) to the attacker.

5.2.2 *Request for Payment*

Detailed analysis of media reports, academic journals, government reports and industry research papers indicates that all known WannaCry ransom requests for payment were sought through the cryptocurrency Bitcoin. This was the sole method and platform currency in which payment was requested. Ransom amounts varied slightly depending on the source of the report with Symantec reporting that requests ranged from $300 to $600 (USD) worth of Bitcoin (Symantec 2017). The United Kingdom's National Audit Office reports indicated the ransom was set at $300 (USD) (Newman 2017), and in the United States, RAND researchers reported "the perpetrators of the attacks demanded a Bitcoin payment of $300 (USD) be deposited in exchange for unlocking the victim's data" (Gerstein 2018).

Investigation of the ransomware requests for payment supports the hypothesis that cryptocurrencies are a primary factor in ransomware attacks. Despite the attackers using only four Bitcoin wallets to collect ransom payments, authorities were unable to prevent or block payments to or from those accounts. Over a year has passed since the global outbreak, with global law enforcement agencies and cybersecurity researchers remaining unable to publicly identify any of the attackers. The DHS has publicly linked the attacks to North Korea or a state-sponsored subsidiary group who ultimately laundered the profits and then transferred them onto the attack's orchestrators (Bossert 2017).

5.2.3 Resolution and Attribution

On the evening of May the 12th, cybersecurity researchers made a breakthrough by successfully identifying and activating an internal kill switch within the WannaCry ransomware source code that was designed as a remote mechanism to stop the attack from propagating any further (National Audit Office 2018). At the time, it was widely reported that a security "researcher using the pseudonym MalwareTech ended up accidentally activating the kill switch when he tried to create a sinkhole in order to study the software."[2] Analysis of the WannaCry program revealed that the source code frequently checked if a specified domain had been registered. Upon received response from the domain, the ransomware was designed to shut down. If no response was received from the domain, the malware continued to spread. Ultimately, when the domain was registered by MalwareTech, this activated the ransomware's internal kill switch (Winckles 2017).

At this point in time, no individual or group has been charged or prosecuted for the WannaCry attacks that impacted over 150 countries. Law enforcement and intelligence agencies in the United States have blamed the attack on North Korean agents, with US Homeland Security Advisor Tom Bossert stating "After careful investigation, the United States is publicly attributing the massive WannaCry cyberattack to North Korea. We do not make this allegation lightly. We do so with evidence, and we do so with partners" (Bossert 2017). Under media questioning about the attack and investigation, Bossert continued, "I will note that, to some degree, we got lucky. In a lot of ways, in the United States we were well-prepared, so it was not luck — it was preparation, it was partnership with private companies, and so forth. But we also had a programmer that was sophisticated, that noticed a glitch in the malware, a kill-switch, and then acted to kill it" (Bossert 2017).

The preparation and partnerships Bossert were referring to were later identified as Microsoft and Facebook. However, analysis of the attacks and these partnerships indicates they had relatively no influence or impact on the spread, management or resolution of the WannaCry attacks. In the wake of the attacks, Facebook reportedly undertook administrative action to close accounts linked to the supposed attackers. Prior to the attacks, Microsoft released a free patch in March 2017 for older versions of Windows software that may have prevented the spread of WannaCry. However, in the wake of the attacks, these systems were already infected, thus installing the patch would have had nil effect in restoring those systems. Kristen Eichensehr (2017), a professor at UCLA School of Law, disputes Bossert's attribution to North Korea:

> The accusation came first in a Wall Street Journal op-ed by U.S. Homeland Security Advisor Tom Bossert... Attribution by op-ed doesn't lend itself to technical detail. Prior U.S. attribu-

[2] Note: In 2018, an FBI investigation in WannaCry identified Marcus Hutchins as MalwareTech. Whilst initially Hutchins was hailed a hero for his role in stopping WannaCry, he was later arrested and has plead guilty for the development of Kronos malware. Kronos was a piece of malware used to steal banking credentials. (See Winder 2019).

> tions, particularly the attribution of the Sony hack to North Korea three years ago, have come in for criticism for providing insufficient detail to support accusations, and this attribution is the least-supported to date (Eichensehr 2017).

Eichensehr's claims are not isolated, with many other cybersecurity experts highlighting amateur-level mistakes in WannaCry's code and execution: "it increasingly appears that this is not the work of hacker masterminds. Instead, cybersecurity investigators see in the recent meltdown a sloppy cybercriminal scheme, one that reveals amateur mistakes at practically every turn" (Greenberg 2018b). In an extensive report into the ransomware attacks, *WIRED* concluded of the many errors the attacker made; the most notable is:

> Building in a web-based 'kill-switch' that cut short its spread, unsavvy handling of bitcoin payments that makes it far easier to track the hacker group's profits, and even a shoddy ransom function in the malware itself. Some analysts say the system makes it impossible for the criminals to know who's paid the ransom and who hasn't (Greenberg 2018b).

WannaCry is not the first cyberattack to be attributed to North Korea. Multiple sources including law enforcement agencies and security researchers have attributed the $80 million (USD) cyber heist from the Bangladeshi Bank and subsequent money laundering of the stolen funds through casinos in the Philippines to North Korean agents (Chow 2018). There are limited hard facts to corroborate these accusations; however, the implementation of tougher United Nations sanctions may be a driving force behind North Korea's interest in cyberattacks. This shadow cyber economy has provided vital support for the regime and the elite who run the country. Emerging technologies such as cryptocurrencies enable an illicit financial system for North Korea to evade international sanctions whilst continuing to proliferate and trade military goods and services (Bechtol 2018).

Internationally, attribution has been mixed. In the United Kingdom, the UK's Government Communications Headquarters (GCHQ) concluded the "attack was not specifically targeted at the NHS and is affecting organisations from across a range of sectors" (National Health Service 2017).

"NHS England's IT team did not have on-call arrangements in place, but staff came in voluntarily to help resolve the issue. Front-line NHS staff adapted to communication challenges and shared information through personal mobile devices, including using the encrypted WhatsApp application. NHS national bodies and trusts told us that this worked well on the day although is not an official communication channel" (National Audit Office 2018, p. 24).

5.2.4 *Results*

Despite WannaCry being one of the largest cyberattacks in history, it appears to date that the proceeds of the attacks only total between $80,000 and $140,000 (USD) worth of Bitcoins (Meyer 2017). This amount is significantly lower than previous ransomware attacks, which infected substantially less systems but were significantly more profitable. Whilst the ransoms paid to the attackers may be far less than anticipated for an attack of this nature and scale, even conservative estimates indicate the attack's fallout cost may exceed $1 billion (USD). As a result of the attacks against the UK's NHS, more than 25,000 medical appointments were either cancelled or rescheduled. The true cost of the attack is yet to be determined with NHS reporting:

> The Department does not know the cost of the disruption to services. Costs include; cancelled appointments, additional IT support provided by local NHS bodies, or IT consultants, or the cost of restoring data and systems affected by the attack. National and local NHS staff worked overtime including over the weekend of 13-14th May to resolve problems and to prevent a fresh wave of organisations being affected by WannaCry on Monday 15 May (National Audit Office 2018, p. 8).

A report in 2018 by *WIRED* found that WannaCry was the most notorious worm-based cyberattack to date, with estimates revealing that globally the attacks had costs between $4 billion and $8 billion (USD). Despite infecting over 300,000 devices and causing widespread chaos across infected system, data analysis indicates very few victims actually paid the ransom. Analysis by researchers at Cornell University revealed that all the Bitcoins in the account linked to the attacks were withdrawn in August 2017, totalling at the time approximately $84,000 (USD) (Conti et al. 2018).

Comprehensive analysis of the WannaCry ransomware attacks indicates that the cyberattacks were reliant on emerging technologies such as advanced encryption techniques, cryptocurrencies and Internet anonymity. The attacks demonstrated a clear ability to monetise both crown jewels and innocuous user data using encryption and extortion techniques.[3] Whilst there remains confusion about the purpose of embedding of an internal kill switch in the attack, this feature severely hampered the profitability of the attacks. As a result, in a relatively small period of time, cybersecurity providers and researchers were able to reverse engineer the attack's source code and limit the severity of future attacks by preventing the encryption sequence.

5.3 NotPetya

The NotPetya attacks began in early June of 2017, with most of the initially infected device belonging to organisations that have commercial offices located in Ukraine. The name NotPetya was derived from the attack's resemblance to an earlier series of ransomware attacks in 2016 dubbed Petya. The timing of the attacks, which struck on the eve of the Ukraine's constitution day (the official day that marks its split from the Soviet Union), has led many security and political analysts to believe the attacks originated in Russia. In the period since the dispute over Crimea erupted,

[3] Note: The term "crown jewels" is a cybersecurity term synonymous with high-value data and systems. The term broadly applies to an organisation's high-value data which typically includes intellectual property, customer data and privileged user account information.

the Ukraine has suffered thousands of well-orchestrated cyberattacks against its financial institutions and critical infrastructure operators.

The NotPetya ransomware attacks hit just months after the WannaCry ransomware outbreak. The attacks built on exploits used in previous ransomware attacks, re-engineering leaked NSA exploits to infiltrate and circumvent the latest security patches from Microsoft. The speed and level of re-engineering led many security researchers to consider the NotPetya attacks to be state sponsored. These considerations have been further fuelled by numerous technical examinations of the ransomware's source code, which indicate that the attacks were not designed for financial gain but for an ulterior motive.

5.3.1 Attack Methodology

Initial reports surfaced, indicating that like WannaCry, NotPetya was being distributed using an email vector. This initial discrepancy most likely occurred because some of the earliest machines infected were already infected with Lokibot malware prior, with those malware infections being already traced to an email vector (Maynor et al. 2019). The NotPetya attack was designed using a worm so that it could be spread automatically, rapidly and indiscriminately. And within a day, the ransomware attack raced beyond the Ukraine, infecting countless computer systems around the world from hospitals in the United States to a chocolate factory in Australia. Technical analysis by the security vendor Crowdstrike discovered:

> NotPetya combines ransomware with the ability to propagate itself across a network. It spreads to Microsoft Windows machines using several propagation methods, including the EternalBlue exploit for the CVE-2017-0144 vulnerability in the SMB service. This is the same vulnerability Microsoft reported on in MS17-010, which was exploited so successfully in the recent WannaCry ransomware outbreak (Sood and Hurley 2017).

The attacks used a virulent mechanism to move laterally spreading from computer to computer. This is a shift in attack vector from most conventional ransomware attacks that are spread through phishing campaigns. Mounir Hadad, a senior director at Cyphort Lab, argues the ability to jump the gap between computers is a technical-level advancement above attacks we have seen to date. He also argues it may inspire the next generation of ransomware attacks (Anderson 2017). This ability to propagate may be a reason that NotPetya was able to spread so rapidly and bring down systems that were designed to be resilient. Many victims complained that the spread was so quick that they were unable to access credit-card payment systems or withdraw cash from ATMs. This left many Ukrainians wondering whether they had enough money for groceries and petrol (gasoline) to last through the attack. They were uncertain about when they would receive their pay cheques and pensions and whether their prescriptions would be filled (Greenberg 2018a).

5.3.2 *Request for Payment*

An examination of widespread media reports, academic journals and industry research papers indicates that all known NotPetya ransom demands for payment were sought through the cryptocurrency Bitcoin. As detailed below in Fig. 5.3, the attackers requested $300 (USD) worth of Bitcoin to be deposited into an account and for the victims to then send a receipt for the transactions to a designated email account. Once this information was received, the attackers would email the victim the decryption key.

The attacks were spreading rapidly around the globe; however, there was a problem. Soon after the attacks were launched, the email account provider disabled the attacker's email account. This meant that any victim that had paid the ransom now had no method to communicate with the attackers. Social media was flooded with posts about the issue, with many experts warning victims not to pay the ransom demands. The disablement of the email account was only the first problem for the victims. As the world's managed service providers were scrambling to restore systems, security researchers were frantically examining and trying to reverse-engineer NotPetya's source code. Within days of the outbreak, security researchers had found something more sinister in the attacks code – the attacks were designed to conceal and destroy data, not to make profit. Numerous security researchers found after analysing the encryption process used in the NotPetya ransomware that "we have determined that the attacker is unable to decrypt the victim's disk, even if payment has been made. These results were based on the source code not extracting or forwarding any of the necessary decryption information to the attacker" (Ivanov and

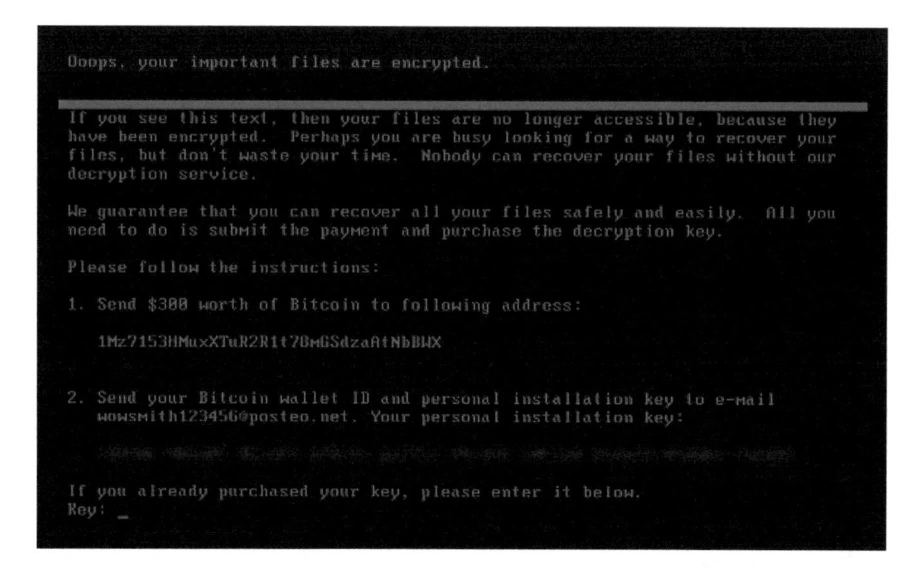

Fig. 5.3 NotPetya ransom demand

Mamedov 2017). This led to widespread pleas from security researchers and vendors not to pay the ransom demands.

A year-long investigation by *WIRED's* Andy Greenberg into NotPetya told the story of absolute destruction. The story told how NotPetya had infiltrated A.P. Møller-Maersk, brining one of the world's largest logistics and maritime companies to an abrupt halt. It's estimated that NotPetya cost Maersk $300 million (USD), and that does not include the downstream losses incurred by port operators, trucking companies and third-party suppliers. Maersk was not the only victim of the attacks, with the French construction company Saint-Gobain losing an estimated $384 million, Reckitt Benckiser also lost an estimated $129 million, but all of these were dwarfed by the pharmaceutical company Merck who lost an estimated $870 million (Greenberg 2018a). This investigation also supported the notion that NotPetya was designed to cause damage, not turn a profit:

> NotPetya's ransom messages were only a ruse: The malware's goal was purely destructive. It irreversibly encrypted computers' master boot records, the deep-seated part of a machine that tells it where to find its own operating system. Any ransom payment that victims tried to make was futile. No key even existed to reorder the scrambled noise of their computer's contents" (Greenberg 2018a).

Whilst this discovery is not the first incidence of destructive ransomware, it has changed the paradigm around major ransomware attacks. The attacks have set the scene for a pivotal legal battle over cyber insurance coverage, which has the potential to create a legal precedence that could cause a ripple effect through the cyber insurance industry (Ralph and Armstrong 2019).

5.3.3 Resolution and Attribution

In early July of 2017, Ukrainian authorities began to prevent the spread after seizing infected servers. The servers belonged to M.E.Doc, which is a software-based tax preparation program. After becoming compromised, the servers were used to spread the ransomware. Security researchers who were part of the investigative team allege that the servers were grossly mismanaged and had not been updated since 2013 (Cimpanu 2017).

Interviews with senior security researchers from the Kiev-based cybersecurity firm ISSP who were initial responders to the NotPetya attacks assert that the attack was "intended not merely for destruction but to act as a clean-up effort. They argue the attackers had months of unfettered access to many of the victims' networks before launching the attacks" (Greenberg 2018a). This level of technical proficiency and the attacks being designed to destroy evidence suggest the attacks were used to conceal espionage or even other criminal activity.

In February 2018, in a coordinated effort by seven countries including the US, the UK and Australian governments, the NotPetya attacks were officially attributed to Russia (Taylor 2018). This attribution is consistent with multiple security researchers and vendors attributing the cyberattacks to Russian actors. Cisco's Craig Williams says, "This was a piece of malware designed to send a political message: If you do business in Ukraine, bad things are going to happen to you" (Greenberg 2018a). The attacks also highlight the potential for ransomware to be used as coercion tool in future non-violent conflicts between nation states.

5.3.4 Results

Despite causing billions in damages, the attacks generated almost no profit. However, unlike previous ransomware attacks that failed to generate significant profits, this failure appears to be somewhat deliberate and not the result of technical deficiencies. The attackers successfully infiltrated user systems and encrypted user data, but the attack's source code was designed in a way that was counterproductive to monetising it. The attackers adopted techniques used commonly by organised criminal syndicates to conceal their true objectives. The attackers used a well-known type of cyberattack, exploited unsecured servers to launch the attack and requested payments (albeit for show) to be made in Bitcoin.

NATO Deputy Assistant General for emerging security challenges, Jamie Shea, argues that the attacks against the Ukraine are not isolated and that the "Ukraine has suffered disruption to its election voting system, train and airline on-line booking, ports, electricity grid, and most recently, the massive elimination of tax and financial accounting data through the NotPetya malware" (Shea 2017).

Whilst is it not possible to determine the development period, coordination and command of the NotPetya attacks, they do highlight the potential partnerships between rogue states and organised cybercrime syndicates. The attacks also highlight the potential for future ransomware attacks to be profitable through disruption effects. These profits could be generated through CaaS/RaaS models, but they could also be generated through stock manipulation. Whilst international markets are complex and subject to multiple forces beyond the scope of this research, high-level analysis indicates that three of the largest NotPetya victims had positive performances in the 12 months leading up to the attacks, which quickly turned to significant downturns. This series of events are detailed throughout Figs. 5.4, 5.5, 5.6, and 5.7:

In the period since the NotPetya attacks, all three companies have significantly depreciated in value, which is contrary to the performance of other publicly listed companies of similar sizes who have on average experienced significant growth and increased market cap values.

Examination of Fig. 5.11 below indicates there could be up to a negative 45 percent deviation in company value (market capitalisation) almost 18 months after the attacks. Based on the market value of these companies, these outcomes indicate that even when consumer trust has not been breached, ransomware attacks may have prolonged financial effects well beyond recovery times. Within Fig. 5.8, the S&P500 is illustrated in red, Maersk in black, Reckitt Benckiser in blue and Saint-Gobain in grey.

Fig. 5.4 A.P. Møller-Maersk stock price for 5 years

Fig. 5.5 Reckitt Benckiser stock price for 5 years

Fig. 5.6 Saint-Gobain stock price for 5 years

Fig. 5.7 S&P 500 index price for 5 years

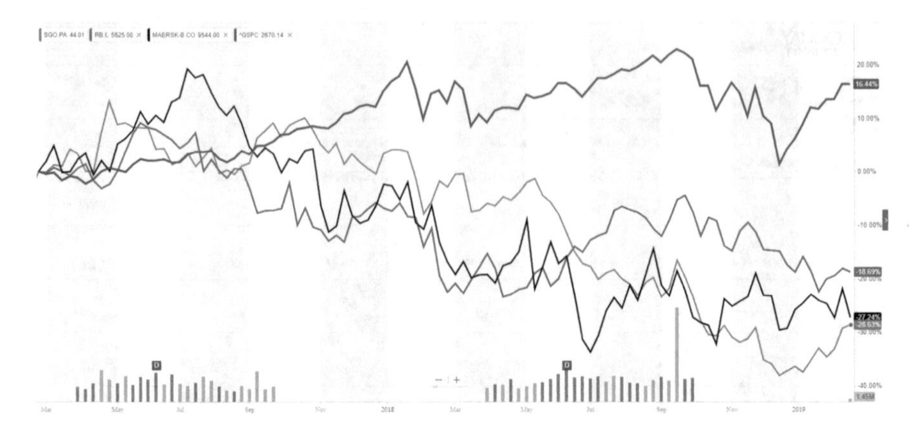

Fig. 5.8 Comparison of stock price for 2 years

5.4 Locky

The Locky ransomware attacks began in February 2016 when the Hollywood Presbyterian Medical Center in Los Angeles, California, became infected with Locky ransomware. Initially, "the infection encrypted systems throughout the facility, locking staff out of computers and electronic records. The attack was eventually concluded when the hospital paid a ransom of 40 Bitcoins ($17,000 USD), in order to acquire the decryption key to restore its data" (Smith 2016). After the attacks, the president of the Hollywood Presbyterian Medical Center Allen Stefanek released a statement stating "the quickest and most efficient way to restore our systems and administrative functions was to pay the ransom and obtain the decryption key. In the best interest of restoring normal operations, we did this" (Smith 2016). However, this wasn't the end of Locky, and by mid-2016, the FBI had released an alert about Locky ransomware. The alert indicated the FBI was observing Locky attacks against organisations in the United States, Australia, Germany, New Zealand and the United Kingdom. The alert emphasised that the ransomware was propagating through phishing emails that included malicious JavaScripts, Microsoft Office documents and/or compressed attachments (Federal Bureau of Investigation 2016).

5.4.1 Attack Methodology

Locky ransomware was distributed through the Necurs botnet, which is a distributed network that has been described as "a zombie army of over five million hacked devices" (Palmer 2017). When not being used to distribute ransomware attacks, the botnet intermittently transitions between being used for other criminal activities and going offline. In 2012, Necurs emerged as "an infector and rootkit, and quickly

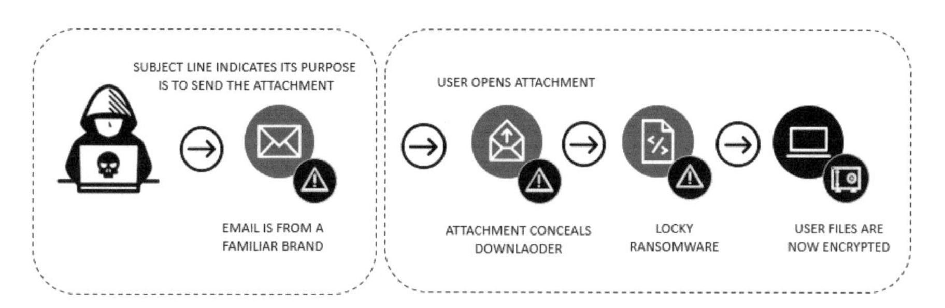

Fig. 5.9 Locky email distribution cycle

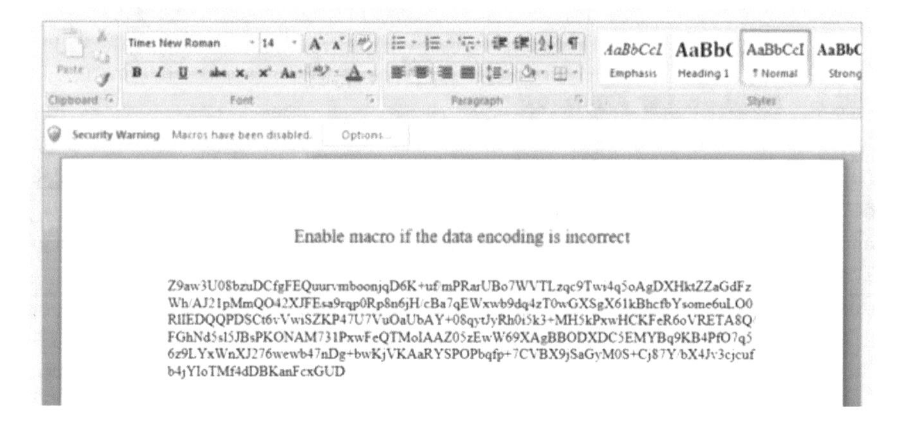

Fig. 5.10 Locky Microsoft Word example

partnered with organised cybercrime syndicates to become part of the top spamming and infection forces in the malware realm" (Kessem 2017). The utilisation of the Necurs botnet has confused security researchers due to the botnet's irregular activity throughout the past 2 years. This intermittent activity cycle is puzzling because to date, it has not been correlated against known law enforcement or social takedown efforts. Another challenge with Necurs is redevelopment. During the periods of inactivity, the Locky ransomware source code has continued to evolve. To make the situation more puzzling, this evolution is often before security researchers have finished disabling the current variant.

The original Necurs botnet distributed millions of emails containing Microsoft Office files. The scale of the email distribution is enormous, with security vendor AppRiver reporting that over 23 million phishing emails were sent containing the Locky ransomware over a 24-hour period (Troy 2017). The attack sequence is illustrated below in Fig. 5.9.

Once the email has been received and opened by the user, it prompts the user to enable macros if the text is undecipherable as detailed below in Fig. 5.10. Enabling the macro initiates a background process which begins to download and encrypt the user's files. Newer variants of Locky have also been concealed in other file types such as excel, zip and even voice messages. Opening the emails attachments mirrors the attack sequence of previous variants, with newer variants of Locky demonstrating greater ferocity in deleting and encrypting recovery options.

Researchers have also identified another alternative attack methodology that uses Facebook messenger to spread Locky ransomware. The ransomware is distributed via a downloader, which bypasses Facebook's inbuilt whitelisting by pretending to be a scalable vector graphic (.svg)-type image file. Facebook has officially denied the platform was exploited to distribute ransomware, instead attributing the spread of ransomware to a defective Chrome browser extension (Ragan 2016). However, the attribution to Chrome is disputed by independent security researcher Peter Kruse (2016).

5.4.2 Request for Payment

Once infected, victims are prompted with a ransom demand. The ransom demand contains details on how to make payment and even included a step-by-step process about how to pay the ransom. Initially, victims are asked to install the anonymising Tor browser before directing the victim to a decryption service to pay the ransom demand. Sophos security researches reported that most Locky victims were being asked for a ransom payment between 0.5 and 1.00 Bitcoin (Ducklin 2016). Based on the market value of Bitcoin at that point in time, a ransom demand of 0.5 Bitcoin equated to approximately $2000 (USD) to retrieve the decryption key (Saarinen 2017).

It was reported by CSO magazine that analysis of the three different Bitcoin accounts associated with the Locky ransomware attacks has generated over $150 million (USD) in revenue (Korolov 2017). Whilst it is difficult to ascertain the true revenue of Locky due to the constant reconfigurations, Google researchers found that Locky was the first ransomware to be earning over $one million (USD) a month, with total ransom payments exceeding $7.8 million (Bursztein et al. 2017). The researchers also noted that they excluded a large number of transactions from the total revenue figures; excluding any transactions that prevented the researcher having the highest degree of confidence, the transaction was a ransomware payment. Based on this payment analysis, the researchers were able to produce the following graph (Fig. 5.11).

5.4.3 Resolution and Attribution

At this point in time, Locky remains relatively dormant; however, its malware variants are not considered to be resolved. Malwarebytes has reported numerous fresh malicious spam campaigns being pushed through the Necurs botnet. Malwarebytes researchers explained the "ups and downs of Locky remain shrouded in mystery.

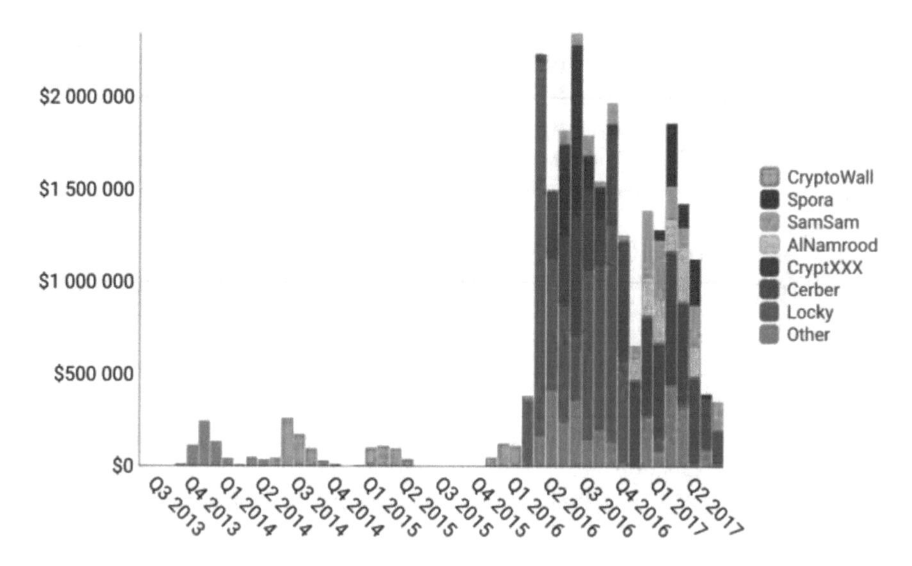

Fig. 5.11 Ransomware payment graph (Bursztein et al. 2017)

One thing time has taught us is that we should never assume Locky is gone simply because it's not active during a specific time period" (Rivero 2017). During these periods of absence, Locky's developers have commonly used this downtime as a chance to build upon their prior successes, developing new and smarter ways to deploy their ransomware.

McAfee lead scientist Christian Beek reported that in Q4 of 2017, their research identified that the Necurs and Gamut botnets together were responsible for 97 percent of the world's botnet spam traffic. Necurs is currently the world's largest spam botnet, distributing 60 percent of the worlds spam botnet traffic. The "infected computers operate in a peer-to-peer model, with limited communication between the nodes and the control servers" (Beek 2018). Beek also found that the botnets controllers advertise the botnet to be hired. This finding also suggests that the Necurs botnet may have been rented for use in the Dridex banking trojan, GlobeImposter, and Scarab ransomware attacks.

5.4.4 Results

At present, the Locky ransomware remains unresolved, and therefore, it continues to pose a significant threat to organisations and individuals. Due to the size of the Necurs botnet, when the botnet is used to spread ransomware attacks, large volumes of Internet users may rapidly become infected with Locky or a variant of the Locky ransomware. Locky has displayed an ability to use emerging technologies and encryption to an attacker's advantage to distribute ransomware and subsequently gain control over a user's Internet-enabled device. Coupled with the use of the cryptocurrency Bitcoin, the attackers have repeatedly demonstrated an ability to monetise valuable and innocuous user data. It has also been argued that resilience of the Necurs botnet may present a larger security threat than the malware it distributes. Threat researchers from IBM investigating the Necurs botnet discovered:

> The most significant point about Necurs is that, unlike common botnet malware such as Kelihos, Necurs has kernel-mode rootkit capabilities. It is composed of a kernel-mode driver and a user-mode component. Another notable point about Necurs is its modular architecture. Typical botnet malware that's only designed to amass bots for other purposes, spread spam or download other malware is relatively simple in technical terms. That is not the case with Necurs. The latter is built in the same way that the most sophisticated Trojans are built, in a modular fashion that allows it to take on different new features as its developers see fit (Kessem 2017).

The botnet is also designed to circumvent Windows firewalls and antivirus software and uses an inbuilt Domain Generation Algorithm (DGA) to generate rendezvous points. In April 2017, Cisco's Talos division attributed the control of the Necurs botnet to organised Russian cybercriminals (Brewster 2017). However, almost 3 years after the initial Locky ransomware attacks, law enforcement agencies have not attributed the attacks to anyone, and there have been no successful prosecutions at this point. Microsoft continues to partner with global law enforcement in a sustained effort to bring down the botnet. Until that can be achieved, many security researchers continue to warn users that Locky remains unresolved and that they should not make any assumptions about the threat being mitigated – it may simply be down for renovations.

5.5 Chapter Conclusions

All four ransomware cases studies analysed demonstrated similar attack vectors but different attack outcomes. Whilst the outcomes are relatively consistent, the attack methodologies share a series of common linkages that have benefitted from the parallel emergence of encryption technologies. All four attacks successfully demonstrated the ability of adversaries to circumvent deployed security controls. All four breaches resulted in the adversary (or malware proxy) successfully deploying advanced encryption techniques against their victims' systems and stored data. Once the encryption process was successfully completed (denying users access to their data), all four adversaries (through automation) subsequently demanded a payment through anonymous cryptocurrencies such as Bitcoin. Interestingly in one case, the ransom demand was used as a deliberate feign to divert attention away from other activities. Holistically, the attacks crippled many of their victim's ability to operate effectively for extended periods of time, with many experiencing significant financial, productivity and reputational losses. Time-series analysis of ransomware attacks since 2016 indicates that the speed of propagation, attack complexity and extended recovery times are significant components why ransomware has emerged as a prodigious cyberthreat to enterprises and governments.

Whilst the success of all four ransomware attacks can be considered mixed from a purely profit-making perspective, the analysis indicates that profit was the underlying motive for all of the attacks. What is clear is that all of the ransomware attacks benefitted from previous cyberattacks. With all four of the ransomware attacks, the adversaries were able to draw on previous attacks, known exploits, and request for payment techniques to further develop these components to their own advantage. This is visible in the source code, the attack speed, attack stealth, payment anonymity and in the overall ingenuity of the ransomware attacks. This further extends the proposition that as encryption technologies have emerged, the general public's ability to anonymously communicate and access information have drastically improved. Subsequently, this aids attackers more than defenders because the majority of defenders have limited motives to anonymously communicate and access information.

Additionally, it was observed that security researchers tend to publicly share their findings on previous ransomware attacks. Logically, attackers could also use this research as a quasi-peer review process to further develop future ransomware attacks. This may provide some correlation to why officials and researchers are continuously stating at the outset of each new major ransomware attack that this attack is the largest, the most sophisticated or the most notorious to date. These comments are clear indicators that ransomware attacks continue to evolve at an agitated pace.

The continued success of ransomware attacks highlights ransomware's continued ability to monetise both high-value and innocuous user data by using computer-enabled encryption and extortion techniques. The continued success of ransomware attacks also raises a growing concern for the potential application of ransomware in warfare. The evolution from cybercrime to warfare is demonstrated in the public attribution of major ransomware attacks to rogue nation states. Ransomware attacks repeatedly caused widespread chaos in hospitals, transport networks, election voting systems, government departments and across other critical infrastructure systems.

The research results also identified how successful ransomware attacks adversely affects the market cap value of large corporations, and in the future, ransomware attacks could be used to manipulate stock markets. The four ransomware attacks collectively exhibit ransomware's ability to rapidly change the cyber risk management environment. The attacks induce fear, chaos and a heightened sense of vulnerability in the victims and large corporations who fear they could be next. From a practical perspective, this vulnerability is warranted because the parallel emergence of technologies has enabled those with limited resources (attackers) to challenge those with vast resources (corporations and governments). Therefore, it can also be argued that this vulnerability is further compounded in a connected world because many executives lack the technical prowess required to fully understand the ransomware threat, let alone defend their organisations. The rapid rise of ransomware means that increasingly, executives are reliant on their staff, consultants, vendors and law enforcement agencies to mitigate and respond to the prodigious cyberthreat.

At this point, it is not possible to determine unequivocally from the open-source evidence available on all four attacks, the intricate differences in the ransomware attack designs, vectors and deployments, to how they relate with their attacker's intended outcomes. For instance, the deployment of the NotPetya attack appears to specifically target Ukrainian businesses, not just random organisations. From a high level, this attack mimicked previous ransomware attacks; however, the targeted design enabled the attack to be executed at maximum speed, which in turn was intended to cause the maximum level of chaos, disruption and damage. The attack utilised previous attacks to conceal the attacker's true intentions, which in turn also delayed and diverted the responder's resources towards paying a fake ransom. This attack highlights the potential relationship between an attacker's motivation and attack design. This reinforces that not all ransomware attacks are random, the result of phishing or some other form of generic attack. If not already, soon we may see ransomware deployed as a coercive tool in a high-stakes game of political brinksmanship.

References

M. Alazab, Profiling and classifying the behavior of malicious codes. J. Syst. Softw. **100**, 91–102 (2015)

R. Anderson, GameOver Zeus botnet disrupted: Collaborative effort among international partners, 7 Nov 2014

M. Anderson, 'NotPetya': Latest ransomware is a warning note from the future, *IEEE Spectrum* (2017). Available online: https://spectrum.ieee.org/tech-talk/computing/it/notpetya-latest-ransomware-is-a-warning-note-from-the-future. Accessed 22 Feb 2019

Australian Tax Office, Scam alerts. (2020). Available online: https://www.ato.gov.au/general/online-services/identity-security/scam-alerts/. Accessed 17 Aug 2020

B. Bechtol, Enabling violence and instability, in *North Korean Military Proliferation in the Middle East and Africa*, vol. 44, (University Press of Kentucky, 2018)

C. Beek, Necurs Botnet leads the world in sending spam traffic, *McAfee Labs*. (11 Mar 2018). Available online: https://securingtomorrow.mcafee.com/other-blogs/mcafee-labs/necurs-botnet-leads-the-world-in-sending-spam-traffic/. Accessed 13 June 2018

Berry, A., J. Homan, R. Eitzman, WannaCry malware profile, *FireEye Threat Research*. (2017). Available online: https://www.fireeye.com/blog/threat-research/2017/05/wannacry-malware-profile.html. Accessed 2 Jan 2019

T. Bossert, Press briefing on the attribution of the WannaCry malware attack to North Korea, 19 Dec 2017

T. Brewster, Google warns ransomware boom scored crooks $2 million a month, *Forbes*. (25 July 2019) 2017 [Online]. Available online: https://www.forbes.com/sites/thomas-brewster/2017/07/25/google-ransomware-multi-million-dollar-business-with-locky-and-cerber/#758974576caf. Accessed 17 Jan 2019

E. Bursztein, K. McRoberts, L. Invernizzi, Tracking desktop ransomware payments, *Black Hat*. Las Vegas, 2017 Google

S. Chow, Hacked: The Bangladesh Bank Heist, *Aljazeera*. (24 May 2018) 2018 [Online]. Available online: https://www.aljazeera.com/programmes/101east/2018/05/hacked-bangladesh-bank-heist-180523070038069.html. Accessed 13 Nov 2018

C. Cimpanu, M.E.Doc software was backdoored 3 times, servers left without updates Since 2013, *Bleeping Computer*. 6 July 2017 (2017)

M. Conti, A. Gangwal, S. Ru, On the economic significance of ransomware campaigns: A bitcoin transactions perspective. Comput. Secur. **79**, 162–189 (2018)

Department of Homeland Security, Alert (TA17-132A): Indicators associated with WannaCry ransomware. (12 May 2017)

P. Ducklin, Ransomware -"Locky" ransomware – what you need to know, *Naked Threats*. (2016). Available online: https://nakedsecurity.sophos.com/2016/02/17/locky-ransomware-what-you-need-to-know/. Accessed 24 Feb 2019

K. Eichensehr, Three questions on the WannaCry attribution to North Korea, *Just Security*. (2017). Available online: https://www.justsecurity.org/49889/questions-wannacry-attribution-north-korea/. Accessed 10 June 2018

N. Etaher, G. Weir, M. Alazab, From ZeuS to Zitmo: Trends in banking malware, in *IEEE International Conference on Trust, Security and Privacy in Computing and Communications*, (Trustcom IEEE, Piscataway, 2015)

Federal Bureau of Investigation, FBI Alert – Identification of ransomware variant called Locky, 11 July 2016

L. Garber, Government officials disrupt two major cyberattack systems. Computer **47**(7), 16–21 (2014)

A. Gazet, Comparative analysis of various ransomware virii. J. Comput. Virol. **6**(1), 77–90 (2010)

D. Gerstein, WannaCry virus: A lesson in global unpreparedness. Available online: https://www.rand.org/blog/2017/05/wannacry-virus-a-lesson-in-global-unpreparedness.html. Accessed 3 June 2018

A. Greenberg, The untold story of NotPetya, the most devastating cyber attack in history, *WIRED*. (2018a). Available online: https://www.wired.com/story/notpetya-cyberattack-ukraine-russia-code-crashed-the-world/. Accessed 23 Jan 2019

A. Greenberg, The WannaCry ransomware hackers made some real ametuer mistakes, *WIRED*. (2018b). Available online: https://www.wired.com/2017/05/wannacry-ransomware-hackers-made-real-amateur-mistakes/. Accessed 5 June 2018

A. Ivanov, O. Mamedov, ExPetr/Petya/NotPetya is a wiper, not ransomware. (2017). Available online: https://securelist.com/expetrpetyanotpetya-is-a-wiper-not-ransomware/78902/. Accessed 14 Dec 2018

K. Jarvis, CryptoLocker ransomware, *Threats & Defenses Threat Analysis*. (2013). Available online: https://www.secureworks.com/research/cryptolocker-ransomware. Accessed 3 Jan 2019

L. Kessem, The Necurs Botnet: A Pandora's box of malicious spam, *IBM Security Intelligence*. (24 Apr 2017). Available online: https://securityintelligence.com/the-necurs-botnet-a-pandoras-box-of-malicious-spam/. Accessed 22 Feb 2019

M. Korolov, Ransomware took in $1 billion in 2016 – improved defenses may not be enough to stem the tide, *CSO*. 5 Jan 2017 2017 [Online]. Available online: https://www.csoonline.com/article/3154714/ransomware-took-in-1-billion-in-2016-improved-defenses-may-not-be-enough-to-stem-the-tide.html. Accessed 11 Feb 2019

P. Kruse, Locky spreading through Facebook. (20 Nov 2016). Available online: https://twitter.com/peterkruse/status/800414481545187328. Accessed 2 Mar 2019

E. Lucas, *Cyberphobia: Identity, Trust, Security and the Internet* (Bloomsbury Publishing, London, 2015)

L. Matthew, Boeing is the latest WannaCry ransomware victim, *Forbes*. (2018). Available online: https://www.forbes.com/sites/leemathews/2018/03/30/boeing-is-the-latest-wannacry-ransomware-victim/#218e8ea96634. Accessed 1 June 2018

D. Maynor, M. Olney, Y. Younan, The medic connection, *Cisco TALOS*. Available online: https://blog.talosintelligence.com/2017/07/the-medoc-connection.html. Accessed 22 Feb 2019

A. McLean, WannaCry reportedly hitting speed cameras in Victoria, *ZDNet*. (2017). Available online: https://www.zdnet.com/article/wannacry-reportedly-hitting-speed-cameras-in-victoria/. Accessed 2 April 2018

A. McNeil, How did the WannaCry ransomworm spread?, *Blog.Malwarebytes.com*. (30 May 2017). Available online: https://blog.malwarebytes.com/cybercrime/2017/05/how-did-wannacry-ransomworm-spread/. Accessed 10 June 2018

D. Meyer, WannaCry ransoms suddenly leave attackers, *Bitcoin Wallets*. (2017). Available online: http://fortune.com/2017/08/03/wannacry-ransom-bitcoin/. Accessed 11 June 2018

M. Molloy, Operation Tovar: The latest attempt to eliminate key botnets, *Threat Research*. (2014). Available online: https://www.fireeye.com/blog/threat-research/2014/07/operation-tovar-the-latest-attempt-to-eliminate-key-botnets.html. Accessed 13 Dec 2018

National Audit Office, *Investigation: WannaCry Cyber Attack and the NHS* (National Audit Office, London, 2018)

National Health Service, Statement on reported NHS cyber-attack, 13 May 2017

L.H. Newman, The ransomware meltdown experts warned about is here, *WIRED*. (2017). Available online: https://www.wired.com/story/eternalblue-leaked-nsa-spy-tool-hacked-world/. Accessed 6 June 2018

L.H. Newman, The leaked NSA spy tool that hacked the world, *WIRED*. (2018). Available online: https://www.wired.com/story/eternalblue-leaked-nsa-spy-tool-hacked-world/. Accessed 6 June 2018

Palisse, A., H. Le Bouder, J.-L. Lanet, C. Le Guernic, A. Legay, Ransomware and the Legacy Crypto API, *The 11th International Conference on Risks and Security of Internet and Systems*. Roscoff, France, 5th–7th September 2016 (Springer, 2016)

D. Palmer, Locky ransomware: Why this menace keeps coming back, *ZDNet*. 7 Sept 2017 (2017) [Online]. Available online: https://www.zdnet.com/article/locky-ransomware-why-this-menace-keeps-coming-back/. Accessed 27 Feb 2019

S. Ragan, Malicious images on Facebook lead to Locky ransomware, *CSO*. (2016). Available online: https://www.csoonline.com/article/3143173/malicious-images-on-facebook-lead-to-locky-ransomware.html. Accessed 14 Feb 2019

O. Ralph, R. Armstrong, Mondelez sues Zurich in test for cyber hack insurance, *Financial Times*. New York, 10 Jan 2019–11 Jan 2019

M. Rivero, Locky ransomware returns to the game with two new flavors. (25 Aug 2017). Available online: https://blog.malwarebytes.com/cybercrime/2017/08/locky-ransomware-returns-to-the-game-with-two-new-flavors/. Accessed 25 Feb 2019

J. Saarinen, Hackers launch massive Locky ransomware campaign, *itNews*. 1 Sept 2017, (2017) [Online]. Available online: https://www.itnews.com.au/news/hackers-launch-massive-locky-ransomware-campaign-472295. Accessed 21 Feb 2019

J. Shea, How is NATO meeting the challenge of cyberspace? *PRISM* **7**(2), 18–29 (2017)

J. Smith, Hospital pays hackers $17,000 in Bitcoins to return computer network, *ZDNet*. 18 Feb 2016 (2016) [Online]. Available online: https://www.zdnet.com/article/hospital-pays-hackers-17000-in-bitcoins-to-return-computer-network/. Accessed 22 Feb 2019

K. Sood, S. Hurley, NotPetya technical analysis – a triple threat: File encryption, MFT encryption, credential theft. 29 June 2017. Available online: https://www.crowdstrike.com/blog/petrwrap-ransomware-technical-analysis-triple-threat-file-encryption-mft-encryption-credential-theft/. Accessed 4 Mar 2019

Symantec, Ransom.WannaCry, (2017). Available online: https://www.symantec.com/security-center/writeup/2017-051310-3522-99. Accessed 7 June 2018

A. Taylor, *NotPetya Malware Attributed*. (16 Feb 2018)

S. Thakkar, Ransomware – Exploring the electronic form of extortion. *Int. J. Sci. Res. Dev.* **2**(10), 123–126 (2014)

G. Troy, Locky ransomware attacks ramp up. 28 Apr 2017. Available online: https://blog.appriver.com/2017/08/locky-ransomware-attacks-increase. Accessed 23 Feb 2019

A. Winckles, Here's how the ransomware attack was stopped – and why it could soon start again, *The Conversation*. (2017). Available online: https://theconversation.com/heres-how-the-ransomware-attack-was-stopped-and-why-it-could-soon-start-again-77745. Accessed 21 Nov 2018

D. Winder, WannaCry Hero Marcus Hutchins pleads guilty to creating banking malware, *Forbes*. 20 Apr 2019 (2019) [Online]. Available online: https://www.forbes.com/sites/daveywinder/2019/04/20/wannacry-hero-marcus-hutchins-pleads-guilty-to-creating-banking-malware/#13f645a4513e. Accessed 23 June 2019

J. Wolff, *You'll See This Message When It Is Too Late: The Legal and Economic Aftermath of Cybersecurity Breaches* (The MIT Press, Cambridge, 2018)

Chapter 6
Dangerous Convergences

The focus of this chapter is to ask what has contributed to us reaching this point. It starts by briefly exploring some key technological advances that enabled the creation of the ransomware phenomenon. These advances and challenges in isolation are not insurmountable; however, as they have begun to converge, the risk have significantly increased. The final sections of the chapter examine risk management frameworks, some of the impacts cyber skills deficiencies have on organisational risk management practices and how as collective these components influence the outlook of future ransomware attacks.

6.1 The Influence of Emerging Technologies

The advent of the Internet has had a profound impact on society, from creating new markets to changing the ways in which we connect with each other. Open-source-applied cryptography is at the crux of many of these advances. With increased access to advanced techniques, this has led to a steady decline in the expertise required to apply and deploy advanced encryption techniques. Consequently, the reduction in required expertise and ease of deployment have also played a fundamental role in the further development of ransomware attacks. Only a few decades ago, many of these techniques and technologies were generally unattainable to the public, were restricted in use to advanced military forces and required specialist equipment and knowledge to apply. However, today these techniques are readily available and free for the general consumer to deploy and exploit. These radical advances in technology have left many corporations and government agencies in the dark and scrambling to adapt. On the other hand, organised cybercrime syndicates have rapidly adopted these techniques and technologies, actively seeking out new

M. Ryan, *Ransomware Revolution: The Rise of a Prodigious Cyber Threat*,
Advances in Information Security 85, https://doi.org/10.1007/978-3-030-66583-8_6

ventures that are beyond the reach of traditional law enforcement efforts. Table 6.1 below illustrates the advent and emergence of new technologies and the subsequent major ransomware attacks:

Table 6.1 Timeline of technologies emerging

Year	Technology occurrence	Encryption occurrence	Ransomware	~Number of IoT devices
1977		RSA encryption created		
1983	Internet created			
1985		Elliptic-Curve Cryptography (ECC) created		
1988	Internet Relay Chat (IRC)			
1989			AIDS Trojan	
1995	Tor becomes open source			
1996		Cryptovirology paper presented		
1998		AES created		
2000		Encryption becomes open source		
2003	Wi-Fi created			
2004	Darknet becomes open source			
2005			GP coder discovered	
2008	Bitcoin launched[a]			
2011	Zero-day and Silk Road markets launched			
2013	Ripple launched	Blackberry enables encrypted communications	Gameover Zeus botnet	
2014	Monero launched	Signal releases encrypted communications		
2015	Ethereum launched			
2016	Zcash launched	WhatsApp and Facebook enable encrypted communications	Locky ransomware	6.3 billion
			Petya ransomware	
2017	Bitcoin Cash launched		WannaCry ransomware	8.4 billion
			NotPetya ransomware	
2018			Ryuk ransomware	11.1 billion

[a]Note: The Bitcoin White Paper was released in 2008 with the newtwork launching in 2009. Additionally, Ethereum was originally conceptualised in 2013, later launching in 2015

6.1.1 Internet-Enabled and Internet-Connected Devices

A driving factor behind the increasing number of ransomware attacks is the growing number of Internet-enabled and interconnected devices. This rise in Internet-connected end point devices has also moved much of the intelligence away from core telecom networks; basic telephone handsets have been replaced by smart devices. Over the past decade, the exponential rise in the number of Internet-enabled devices is staggering. Gartner reported that there were 8.4 billion Internet-enabled devices in use globally in 2017, and this is estimated to grow to over 20 billion by 2020 (Van der Meulen 2017). This meteoric rise in Internet-enabled device numbers has also brought with it an increased attack surface. However, this does not suggest that only Internet-enabled devices are vulnerable to ransomware attacks.[1] With more devices connected, future attacks may have increased velocity and impact, simply because there are more opportunities for cybercriminals to deploy ransomware attacks.

Whilst it was not developed to measure the value of Internet devices, Metcalfe's law offers insight into the potential value of clusters of IoT devices and enterprise networks to ransomware attackers. Metcalfe's law states "that the value of a network is proportional to the square of its size, relying on the observation that for a network with n members, each can make n – 1 connection with the other members" (Metcalfe 1995). Madureira et al. (2013) expands on Metcalf's law to detail:

> If all those connections are equally valuable, the total value of the network is proportional to $n\ (n - 1)$, thus roughly to n^2. For example, if a network has 5 members, there are 20 different possible connections that members can make to each other. If the network doubles its size to 10 members, then the number of connections does not simply double, but roughly quadruples to 90 (Madureira et al. 2013).

Applying Metcalfe's law to determine the risk to large enterprise networks from ransomware indicates that the risk exponentially increases with increases to the networks' size. For enterprises to maintain adequate security, security resources must increase proportionally with enterprise device numbers and where necessary reflect the increased level of network complexity.

Another reason why the attack surface continues to expand is the accessibility and functionality of these Internet-enabled devices. Mobile phones are an excellent example of Internet-enabled devices that now utilise operating systems that integrate with Wi-Fi, Bluetooth, Internet browsing, email access and the installation of a vast array of new applications. Concurrently, the emergence and use of these

[1] Note: There is a growing body of evidence that indicates IT systems and devices that are not Internet enabled may be at a heightened risk from ransomware. This heightened risk may be the result of reduced security controls and practices being implemented on these systems due to them being perceived to be at a lower risk. This may include reductions in security controls such as antivirus protection, firewalls, network segregation, patch management and penetration testing.

additional features and functions have significantly increased the size of each individual device's attack surface, which inherently increases the risk of a user's device falling victim to ransomware.

As discussed in earlier chapters, numerous security experts and researchers have concerns that in the foreseeable future, ransomware attacks may begin specifically targeting devices beyond personal computers and popular low-cost consumer IoT devices such as consumer electronics (Barret 2020). The obvious areas for concern are Internet-enabled and Internet-connected devices embedded in critical infrastructure such as electricity, water, communications and transport networks. These security concerns are heightened because of the increased connectivity between IT and OT systems in critical infrastructure environments. This connectivity is being driven by the need for quantitative management reporting to enhance productivity/production efficiency. This desire for efficiency has led to increases in automation, sensors for data analytics, cloud-based operations and remote-controlled operations, creating linkages between IT and OT environments. Whilst this connectivity is inevitably due to other forces, this is problematic in terms of cybersecurity (Murray et al. 2017).

Another emerging sector of concern is the growing industry of autonomous air-, sea- and land-operated vehicles. Whilst these market sectors and devices will undoubtedly continue to face significant security challenges, emerging research suggests that the medical industry may be at the greatest risk. There's a growing awareness among doctors and healthcare providers that they rely on hundreds of devices – the crash carts, insulin pumps, heart monitors, healthcare records and other machines that are integral to patient care.[2] These concerns are not unfounded; when the WannaCry ransomware attack hit, it also infected numerous impacted medical imaging devices, causing disruption to numerous hospitals and medical centres.

The exponential growth of Internet-enabled device numbers is only one component of the broader security challenge. This security challenge is being made substantially more difficult due to the number of Internet-enabled device developers, with *Forbes* estimating in 2017, there were 4.7 million IoT developers who can produce devices (Kroll 2018). The speed of IoT development, lack of security controls and speed of user adoption suggest that ransomware will continue to pose a formidable cyberthreat in the coming decade.

As illustrated in Fig. 6.1, the gap between the number of devices and skilled people continues to grow. Automation has already reduced the volume of manual tasks being undertaken to cybersecurity staff; however, currently, the speed of its development and implementation does not outpace the speed of devices being introduced into global circulation. Automation, data science, Machine Learning (ML), and Artificial Intelligence (AI) technologies are quickly becoming core capabilities for enterprise trying to solve complex real-world problems. They offer hope in reducing the threat from cyberattacks – but they just as easily may exacerbate the threat.

---

[2] A. Coronado in (Allen 2015).

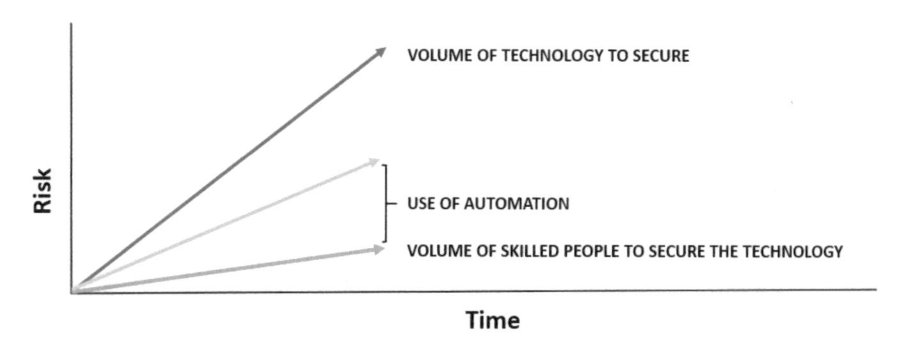

Fig. 6.1 Volume of technology versus skills

6.2 Risk Management

In the interconnected world of modern business and government services, effectively managing cyber risk has quickly become a necessity to achieve and maintain prosperity. Despite continued cyberattacks, many organisations remain underprepared to effectively manage or mitigate the cyberthreats they will be required to face. The Australian government states that "cybercrime will continue to be an attractive option for criminals due to its ability to generate large profits with a low risk of identification and interdiction. Each successful compromise encourages further cybercrime activities" (Australian Cyber Security Centre 2017, p. 15). The modern world is driven by the speed, flow, use and control of data. This growing reliance on data for essential business services means that without adequate cyber risk management professionals to prepare and respond to ransomware attacks, ransomware will continue to present a significant threat to government agencies, enterprises and individual Internet users. This section identifies some of the risk-related factors that influenced how we came to this point and the influence they are having.

6.2.1 Risk Management Frameworks

Within Australia, arguably, the most well-known and commonly applied risk management standard is the International Organization for Standardization 31,000 Risk Management – Principles and Guidelines (ISO31000) (Australian/New Zealand Standards 2009). The standard forms the fundamental basis for risk management for many Australian government departments and large commercial organisations. The standard was originally derived from AS/NZS 4360:2004 Risk Management (Australian/New Zealand Standards 2004). According to Gjerdrum and Salen (2010), it is the first international standard on the practice of risk management (Gjerdrum and Salen 2010). The ISO31000 standard details a generic high-level risk management process (tool) for the management of risk of the activity, discipline

Table 6.2 Example of risk calculator (Australian/New Zealand Standards 2009)

Likelihood / Consequence	Minimal	Minor	Moderate	Major	Catastrophic
Almost certain	Low	Medium	High	Extreme	Extreme
Likely	Low	Medium	High	High	Extreme
Possible	Low	Medium	Medium	High	High
Unlikely	Low	Low	Medium	Medium	High
Rare	Low	Low	Medium	Medium	High

or business sector. The standard has been commonly applied to evaluate and manage risk in sectors such as defence, critical infrastructure, finance and, in more recent periods, cybersecurity.

The process described in ISO31000 for managing risk is identical to that in AS/NZS 4360, incorporating the five steps in the traditional risk management process: identify, analyse, response (select the best option practically available), implement and monitor (Australian/New Zealand Standards 2009). The popularity of the ISO31000 process can be attributed to its simplicity, the ability for it to be broadly applied to a variety of risk events, scalability and the flexibility of the risk analysis process. This allows government and commercial organisations to apply one generic framework to manage a diverse range of non-financial risks across multiple business units and sectors. The following risk calculator provides a common example application of the standard ISO 31000 (Table 6.2):

Cohen's (2017) research indicates that "the dynamic speed of change and the compression of time in cybersecurity move individuals and organisations out of their comfort zones. This often results in forcing faulty decision-making generated by an enhanced dependence on untested assumptions" (Cohen 2017, p. 45). This sentiment is supported by Scheferman (2016), a hacker-turned director for leading cybersecurity vendor Cylance, who explains "as individuals and as a collective society, we are basically novices when it comes to understanding cyber risks, being able to identify an attack, and preparing ourselves for a compromise" (Scheferman 2016).

Two key problems that can be observed from the application of generic risk management frameworks (inbuilt flexibility) are bias and the manipulation of risk calculations. The inbuilt flexibility allows conscious and unconscious bias of the risk analysis process; as a result, the risk analysis outcomes can be easily manipulated to suit the assessor or the organisations desired outcomes.[3] This bias and manipulation are commonly interpreted, described or attributed to the risk analysis process being considered subjective. Hansson (2010) explains "in studies of risk perception the 'objective risk' thus defined is contrasted with the subject's ranking of risk factors that is said to express 'subjective risk' or 'perceived risk'" (Hansson 2010, p. 232).

[3] Note: The categorisation of a risk level may trigger an escalation. Alternatively, a risk reduction may trigger the withdraw of project funding.

Another primary lure of these simple risk calculators' format is the ease in which both the probability and consequences of a specific risk sector can be tailored to meet the client or an organisation's individual needs. For example, the financial value of a consequence can be altered to suit the specific needs of an organisation. Because of this tailoring (manipulation), the benchmark of which defines a Minimal through Catastrophic risk event outcome can be adjusted and applied to a variety of threat types in a variety of areas.

Since the inception of ISO31000, many organisations have adapted the standard's risk calculator (matrix) to simplify or increase the complexity of the generic calculator to suit their own organisational requirements. These adaptions can be further compounded by the assessor deriving inaccurate probabilities. This inaccuracy was demonstrated by economics Nobel Laureate Daniel Kahneman (2011) whose research demonstrated "people overestimate the probabilities of unlikely events and people overweight the probabilities of unlikely events in their decisions" (Kahneman 2011, p. 324).

The adoption and combination of these process failures highlight that cyber risk management frameworks play an important role in the ultimate success of ransomware attacks. Systems and networks whose controls were derived using inaccurate data may negatively impact the risk management process by allocating finite resources to the wrong systems. Whilst it can be argued that this approach may equally lead to both increased and decreased risk control measures, the allocation of finite resources to inconsequential systems may cause critical systems to operate at a heightened risk exposure, thus increasing the organisation's overall residual cyber risk. From a practical perspective, organisations and governments have finite resources to manage and take actions to reduce cyber risk. Inaccurate reporting of cyber risk may also cause inconsequential systems to be prioritised over critical systems.

Currently, in the United States, there is a push to further develop the National Institute of Standards and Technology (NIST) 800 series – Information System Risk Management Framework (National Institute of Standards and Technology 2018). The NIST framework is a comprehensive framework that outlines a series of processes and requirements to achieve a minimum standard of information and cybersecurity across a diverse array of systems and networks. The risk management framework illustrated in Fig. 6.2 provides descriptions of the processes and resources required to undertake each step. This is underlying process that underpins the common mantra "identify, protect, detect, respond, and recover" that is commonly associated with using the NIST800 framework.

Within Australia, the framework is rapidly being adopted by Australian government agencies and enterprises to uplift their cybersecurity posture and management processes. The framework provides a comprehensive step-by-step process, but it does have some limitations. One example limitation may be an enterprise undertaking an upgrade or implementation of a new cybersecurity control. The standard advises that critical systems should have protective and defensive security controls such as firewalls and Intrusion Detection Systems (IDS); however, limited guidance is provided about the implementation or testing process required. As a result, an

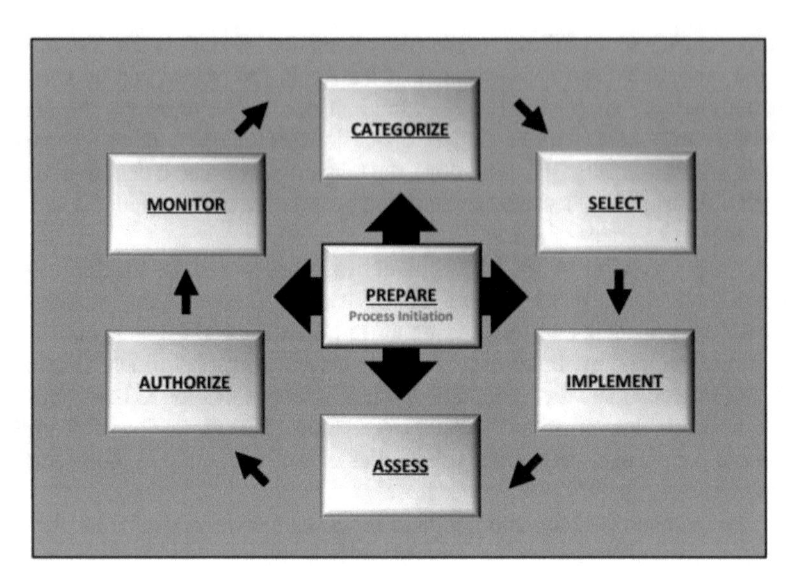

Fig. 6.2 NIST risk management framework

organisation may be compliant with the NIST framework[4]; however, in practice, the security controls may be of limited effectiveness. This is because there is not a defined testing procedure or grading process; therefore, it is difficult to evaluate the control's true effectiveness or level of protection provided. This has the potential to create a scenario where a control may achieve its design effectiveness but not operating effectiveness. To alleviate this potential deficiency, on commissioning (and regularly), all controls should be tested in isolation, as well as being tested using end-to-end scenarios such as those described in the Mitre ATT&CK framework (Strom et al. 2018).

6.3 Cyber Skills Deficiencies

Globally, there is a burgeoning body of academic research indicating and predicting widespread technological and cybersecurity skills deficiencies. Within Australia, these growing academic concerns are echoed by cybersecurity providers, government agencies and business leaders who are becoming increasingly vocal in their concerns about the bleak outlook of Australia's current and future cyber workforce.

[4]Note: Currently, there are no official NIST800 cybersecurity certifiers for private enterprises. Whilst there are no shortage of consultancies that could undertake a gap analysis of enterprises systems, this is not an offficial NIST800 certification of accreditation process. It should also be noted that within US federal agencies, there are no mimimum experience or information system assurance qualfification requirements for the authrorised senior agency official to accept the agency's informations systems risks.

Cisco's Australian vice president Ken Boal has expressed that "there is a major talent gap, not just in technical skills but also the ability to translate the technical situation into the business impact and convert it into meaningful metrics and scenarios for business leaders."[5] These technology and cyber skills shortages are compounded by a new Australian temporary migration policy. The founder of Atlassian, Mike Cannon-Brookes, questions why the long-term temporary visa list includes horse trainers when ICT managers are not on the long-term list.[6]

These widespread skills deficiencies ultimately drive enterprises and government agencies to charge risk committees and individual managers with managing technology and cybersecurity-associated risks they do not comprehensively understand. A report by Information Systems Audit and Control Association (ISACA) in 2017 found that less than 25% of cybersecurity job candidates are qualified for the role (ISACA 2016). This skills deficiency is coupled with reactive processes. Too often despite knowing the risks, organisations continue to adopt reactive cybersecurity management policies that wait for cyberattacks to occur before designating responsibilities, containment plans and the allocation of resources to address the threat. As a result, frequently, it is too late, and the damage to the organisation's reputation and infrastructure has already been done. The quantity and ferocity of cyberattacks, specifically ransomware attacks, indicate there is an underlying cybersecurity risk management problem. Many of these failings are highlighted in United Kingdom's (UK) National Health Service (NHS) processes in the lead up to the WannaCry ransomware attacks on the department.

In the United Kingdom, the NHS is a large government healthcare organisation responsible for treating millions of patients every year. Despite operating in a sector that was considered by authorities to be high risk for targeted cyberattacks and having access to the United Kingdom's advanced cybersecurity agencies and resources, the maturity and security of the systems in use were inadequate. The system's security deficiencies were well known because in the months before the WannaCry ransomware attacks, the "NHS Digital offered an on-site inspection to hospitals to assess their cyber-security (known as 'Care CERT Assure'). This inspection was voluntary, and by 12 May 2017, NHS Digital had inspected 88 out of 236 trusts and none had passed" (National Audit Office 2018, p. 19).

This was not the only risk management failing WannaCry would highlight. Before the WannaCry attacks, NHS had developed an incident response plan for responding to a major cyberattack. The plan detailed the roles and responsibilities across local and national organisations. The problem was NHS had never tested the plan at the local level. As a result, at the outset of WannaCry, the NHS and its local staff were uncertain how to respond and what actions they should be taking (National Audit Office 2018, p. 24). These systemic security and risk management failings are not isolated geographically or to the business sector.

[5] Ken Boal cited in Warner (2018).

[6] Mike Cannon-Brookes cited in Nott (2018).

Looking to the future, this raises a series of underlying questions: Is the problem being compounded by complacency, or do organisations remain sceptical about the actual risks posed by cyber threats? Are organisations reliant on cybersecurity consultants and external security service providers? Analysis of risk management approaches indicates that many cybersecurity frameworks have largely been duplicated from broader risk management frameworks, with minor terminology alterations to suit cyber risk management applications.

6.3.1 Cybersecurity Leaders

For most people in the developed world, they would generally hesitate or be resistive to undergo dental surgery by an unqualified dentist. Learning that your local dentist is not a dentist but instead a mechanic may cause the local community to become extremely anxious and concerned about the risk of undergoing dental surgery with an unqualified dentist. The same community fear and nervousness should be applied to those who manage an organisation's information and cybersecurity. When boards, executives and risk managers are not adequately qualified or experienced for their roles, it is not a question of if they will fail, but when. When these risk managers fail, they may not kill someone, but they will certainly negatively impact their respective organisation's reputation, brand, profits, consumer trust and business longevity.[7]

Rudimentary analysis of the Chief Information Officer (CIO) and Chief Information Security Officer (CISO) or equivalent roles within Australian Stock Exchange (ASX) top 100 firms indicates these roles are commonly filled by persons with varying levels of experience and skill sets. There are further departures when analysing the career paths of these people prior to their appointment in senior information and cybersecurity management roles. This analysis indicates that numerous C-level technology and security leaders have opted for a career pathway that is primarily experience based, undertaking a series of similar roles leading to their position. Few have completed multiple professional cybersecurity certifications, and even fewer have completed formal degrees in fields such as security, risk management, computer science or cybersecurity. This rudimentary analysis suggests that no clear or formal pathway has been established within Australia's largest enterprises and agencies to reach the most senior information, technology and cybersecurity management roles.

Whilst degrees do not provide cybersecurity surety, they could provide organisations an initial skills benchmark. This blurred pathway to becoming a CIO or CISO is coupled with a general lack of understanding by organisations that have been the basis for some individuals to be appointed to senior management positions they may

[7]Note: It is foreseeable in the future that a death may be caused by a cyberattack. For example, if this event was to occur on a commercial passenger jet, questions would inevitably be asked about the cybersecurity practices of the aircraft operator and manufacturer.

not be qualified or experienced enough to adequately fulfil. In the wake of the Equifax breach, where 150 million American consumer records were stolen, it was revealed that Equifax's CISO had not completed a security-related degree or any formal security, risk management, technology, information or cybersecurity training (Popken 2017). Many cybersecurity experts may challenge the validity or requirement to undertake a degree or similar formal training, citing their own path as evidence that this is not required. This may have been and continues to be an acceptable approach for some organisations in the short-term future; however, as the cybersecurity field matures, it can naturally be expected that the minimum training and education requirements for senior leadership positions will increase.

To further demonstrate the uncertainty in cybersecurity training and the pathway to senior management positions, comparisons should be made with the pathways of other professional disciplines. For example, should an Australian student in 2019 wish to become a cardiologist, then they must initially complete an undergraduate degree (i.e., Bachelor of Arts, Commerce, Science, etc.), they must then complete a Doctorate in Medicine with 2 years residency in a public hospital. From this point, doctors interested in becoming a cardiologist can apply – with no guarantees – to the Royal College of General Practitioners to undertake a fellowship that generally takes 6 years to complete (The Universityof New South Wales 2018). Once this is completed, they will be awarded FRACP and, from this point forward, be considered a cardiologist. Whilst there are few similarities between the academic skills and practical experiences required to become a cardiologist and those required to become a CIO or CISO, it does highlight the infancy of cybersecurity discourse and the challenges organisations face in finding qualified and experienced cybersecurity leaders and board members.

The comparison between cybersecurity and Western medicine, a field which has some 3000 years of practical experience coupled with approximately 400 years of scientific experiment to draw on, highlights the infancy of the cybersecurity field. To begin to solving challenges such as ransomware, cybersecurity is going to need lure intelligent and innovative students that in previous generations may have opted for alternative career paths. Overcoming adversity requires ingenuity and strong leadership, and this is absent from many organisations at the upper echelons. The creator of the virtual CISO concept Barak Engel (2018) explains:

> In my 20 plus years of adventures working, learning, and contributing to both the information security and intelligence communities, I can still say I have only met a handful of noteworthy leaders. I guess that's to be expected since they say that true leadership is quite rare…but should it be? In the information security field, it seems more than rare. Security leaders seem almost non-existent (Engel 2018, p. IX).

Barak continued this sentiment explaining that there are two challenges: "the first is the mistaken notion that security problems are generally solved via technology, and the second notion is that security is part of IT in the first place. It shouldn't be" (Engel 2018, p. 8). The ability to manage and protect large enterprises from ransomware attacks requires more leaders to be developed who are qualified and experienced cyber and technology risk managers, coupled with a strong business acumen. Cybersecurity is an expensive overhead that continues to increase for enterprises. Enterprises should be careful of leaders that push aggressive new strategies such as cloud adoption and SaaS solutions, that do not understand the cost of true cost or business benefits.

As a consultant and auditor, I commonly encountered senior leaders who were purely focused on adopting cloud-based solutions. Whilst there are numerous benefits from cloud-based solutions such as security, resilience, speed, scaling, monitoring, etc., there can be numerous disadvantages too. I frequently reviewed technology solutions that were trapped in this quasi hybrid model.[8] The enterprise had moved or adopted a new solution and, however, was unable or did not want to move all its applications or infrastructure to the cloud. As a result, the enterprise was required to deploy a series of controls and tools for both the on-premise and cloud environments. This increased the total number of controls and tools in operation, which also increased the complexity of the environment.

The hybrid model is common for organisations at the outset of their cloud journey. Many foresee this hybrid model as a temporary step before adopting a fully cloud-based environment. The problem is that I have observed very few organisations successfully develop a strategic business plan with sufficient detail to implement the shift, let alone execute it. This does not mean that all hybrid models are more expensive, but it does serve as a warning that complexity equals cost, and switching to the cloud does not guarantee improved security or lower operating cost.

6.4 Practical Risk Management

Cyber experts continue to tout that by following a few key steps, consumers can create resilient networks that can defend and recover from ransomware attacks. This simplified portrayal may be achievable for some individual users and small-to-medium businesses; however, those experienced in the change management process inside large enterprises and government networks will paint a very different picture. Industry sectors such as government services and financial services have rapidly implemented new technology schemes to deliver faster and more integrated services. Despite being able to rapidly implement these technologies, many of these organisations continue to operate large fleets of legacy systems. There are a variety of commonly cited reasons underlying why legacy systems have not been replaced including the system is not broken, we can't take that system down, the cost of upgrading is too high and the network is too complex. Although these are legitimate reasons, they have the potential to adversely impact the security posture of an organisation by complicating the change management, vulnerability management and security patching process.

[8] Note: When discussing cloud infrastructure, the term hybrid commonly refers to the infrastructure environment being split between on-premise and cloud hosted (externally operated data centre) environments.

In complex organisations, it may not be possible to accurately identify, remove, mitigate or transfer all known cyberthreats and cyberattack vectors. However, these limitations need to be documented and conveyed to regulators, management and the board so that conscious decisions can be made by those charged with managing the residual risks. In the wake of a ransomware attack, it is vital that key staff understand their roles and responsibilities and the resources available and not available. In the WannaCry attacks against the UK's NHS, the "NHS had not rehearsed for a national cyber-attack, as a result it was not immediately clear who should lead the response and there were problems with communications" (National Audit Office 2018, p. 9). Failure to undertake adequate training and preparation only inhibits an organisation's ability to detect, contain and respond to ransomware attacks, which is detrimental to customers, staff, operating costs and, ultimately, the organisation.

6.4.1 External Risk

There are numerous cyber-associated risks that are considered beyond the control of individuals and business organisations. At the top of the threat list are zero-day exploits. Many cybersecurity experts and researchers consider these zero-day exploits to present the most dangerous threat to an organisation's security. A classic example of external risk is the deliberate insertion or ongoing concealment of vulnerabilities by software developers and intelligence agencies. Within the intelligence and cybersecurity communities, there remains a long-standing debate about the dangers associated with installing bugs in commercial products for intelligence gathering. There is a growing body of evidence indicating that "agencies keep these flaws to themselves, instead of notifying the company that makes the software, so the vendor can patch the vulnerabilities and protect its customers. If these tools get out, they potentially endanger billions of software users" (Newman 2017).

One example of this occurring is the leaked NSA exploit EternalBlue, which was subsequently used by cybercriminals to undertake ransomware attacks and a variety of other cyberattacks (Newman 2018). However, vulnerabilities are not limited to software applications. In recent years, multiple hardware manufacturers such as Huawei and Hikvision have also been accused of deliberately undermining hardware security. These allegations have become an increasing concern for next-generation telecommunications and critical infrastructure control systems, which has led to numerous hardware manufacturers being excluded from large infrastructure projects and consumer markets.

The concealment of vulnerabilities prevents developers, cybersecurity vendors, risk managers and users from taking the necessary steps to reduce or mitigate the potential threats. This threat is further magnified once the exploit is released into the virtual wild because cybercriminals often quickly modify existing malware to utilise the new exploit before patches can be developed and implemented. An empirical study into zero-day attacks by researchers at Symantec found that "after the disclosure of zero-day vulnerabilities, the volume of attacks exploiting them increases by up to 5 orders of magnitude" (Bilge and Dumitras 2012). This period of heightened vulnerability is supported by the Australian Signals Directorate (ASD), which recommends that all patches should be implemented to all computers systems at risk within 48 hours of their release (Australian Signals Directorate 2020).

6.4.2 Network Containment and Segregation

The continued pursuit and development of zero-day exploits means for the foreseeable future, there will be no silver bullet to preventing ransomware attacks. As a result, the containment of attacks should be at the forefront of those charged with managing cyber risk, designing security controls and responding to ransomware attacks. Containment is a form of strategy that aims to limit an organisation's exposure to cyber intrusions by preventing, reducing or slowing an attacker (or piece of malware) ability to propagate throughout a network. By design, containment is usually achieved using some form of network segmentation and/or segregation. The ASD explains that "network segmentation involves partitioning a network into smaller networks; while network segregation involves developing and enforcing a ruleset for controlling the communications between specific hosts and services" (Austalian Government 2019). The ASD also advises organisations that network segregation should be applied across multiple layers of the network and, for sensitive environments where possible, should include physical segregation (air gapping) of networks and hosts.

Another crucial element in containment strategies is the deployment of firewalls and intrusion detection notification systems. Firewalls are an integral security control measure; however, simply zoning networks and deploying firewalls at external access points have repeatedly been proven to be ineffective at preventing or containing ransomware attacks.[9] Firewalls should be deployed both at internal and external access points and, at a minimum, should be supplemented with IDS and Intrusion Protection Systems (IPS) that can help monitor unwanted activity across the network. Enterprises should also regularly undertake both external and internal penetration testing. Many organisations fail to undertake internal penetration testing, instead focusing only on the external threats. Completing internal penetration testing is essential to inhibiting an attacker's ability to move laterally and for containing self-propagating ransomware. The deployment and testing of these security controls in conjunction with application whitelisting may provide the most practical methods to contain known ransomware attacks in real time.[10]

[9] Note. This process also tends to create 'flat-networks' which may limit the ability of organisations to apply the principle of defence-in-depth. Commonly flat network architectures are not effective at protecting organisations from contemporary cyber threats.

[10] Note. Patching and scanning are also essential security controls to defend against ransomware attacks.

6.4.3 *Small-to-Medium Enterprises and Individuals*

The rise of the Internet has enabled individual users and small-to-medium enterprises (SME) businesses to undertake business in global marketplaces. Cyber risk management for individual users and SMEs can be a conflicting process. On one hand, these types of businesses and users typically have smaller and less complicated networks where changes can be applied rapidly. On the other hand, they have limited resources and expertise to draw upon to manage their cyber risk. Whilst individuals and SMEs have been rapid adopters of new technologies, this has frequently left their data, networks and systems vulnerable.

The process of establishing and maintaining good cybersecurity and hygiene can be complex and expensive process for individuals and SMEs. Government agencies and enterprises often have large numbers of users and end point devices (which increases their attack surface), but they are also better resourced to draw on specialist cybersecurity expertise. This expertise often comes in the form of annual cybersecurity staff training, specialist IT staff and third parties performing audits, assurance, regular backups and vulnerability management. The basic ransomware defences that individuals and SMEs should be implementing is implementing antivirus (heuristic and behavioural), keeping all software update, making regular offline backups, disabling Microsoft Office macros, using strong passwords (passphrases) and being cautious when opening emails and links.

Ransomware attacks against government agencies and large corporations have the capacity to reap large rewards. However, ransomware attacks against government agencies and large corporations also draw considerable attention from law enforcement and cybersecurity researchers. This attention enables the organisation to draw upon substantial resources, pitting the attacker against well-resourced and powerful organisations. In comparison, ransomware attacks against SMEs and individual users are unlikely to draw the attention of law enforcement. In fact, many victims may not even inform law enforcement of the incident. The speed and complexity of the attack may overwhelm the victim, causing them to enter a state of shock. This overwhelming feeling of helplessness may swiftly provoke the victim to pay the ransom demands. For these reasons, it can be argued that SMEs and individual users are more likely to become victims of ransomware attacks and to ultimately pay the ransom.

6.5 Conclusions

The explosion in number of Internet-enabled devices and IOT devices in use continues to exponentially increase the potential number of victims for ransomware attacks. Smart end point devices have transformed how we communicate, undertake business and undertake our daily lives. However, the continued push to intelligent end point devices has also significantly increased each device's potential attack surface. This combination of an increasing market size (volume of devices) and increasing number of potential attack vectors for each device is good news for cybercriminals and ransomware developers.

In my professional experience, the NIST800 series is rapidly becoming the cyber risk framework of choice for enterprises and government agencies. Whilst this represents a significant improvement over the adaption of generic risk management framework, deficiencies remain within NIST's framework relating to control testing processes. As the NIST800 framework continues to evolve, it is hoped that future versions will address the minimum testing expertise required for controls testing assessors and provide further details about the prescribed testing processes required to determine that the controls are operating effectively.

Rectifying this will be a challenging task because many enterprises lack the required number of appropriately qualified and experienced personnel. This deficiency is evident at the highest levels of business with many technology and cyber leaderships positions filled by people with limited or no formal qualifications in the associated fields. Whilst the public profile of technology and cybersecurity risks continue to increase, enterprise cybersecurity budgets will continue to rise in an attempt to address these risks, and this should also begin to lift the minimum experience and qualification requirements for board and senior leadership positions.

Whilst it is difficult to obtain sufficient data, it is also a logical hypothesis that smaller organisations and individual users are more likely to become the victim of a ransomware attack. This may or may not be a deliberate design objective of the ransomware attack but simply a result of smaller organisations having limited cyber resources. With limited specialist cyber resources at their disposal, many small organisations and individuals will find designing, operating and maintaining a secure network a challenging process.

References

A. Allen, Billions to install, now billions to protect, *Politico*, 1 June 2015 (2015)

Australian Cyber Security Centre, Australian Cyber Security Centre Threat Report. (Australian Government, 2017). Available online: https://www.acsc.gov.au/publications/ACSC_Threat_Report_2017.pdf. Accessed 6 Feb 2019

Australian Government, *Implementing Network Segmentation and Segregation* (A. S. Directorate, Canberra, 2019). Available online: https://www.cyber.gov.au/sites/default/files/2019-05/PROTECT%20-%20Implementing%20Network%20Segmentation%20and%20Segregation%20%28April%202019%29.pdf. Accessed 11 Aug 2019

Australian Signals Directorate, Essential Eight Explained (Department of Defence, Canberra, 2020). Available online: https://www.cyber.gov.au/acsc/view-all-content/publications/essential-eight-explained. Accessed 22 Jan 2021

Australian/New Zealand Standards, *AS/NZS 4360: 2004 Risk Management* (Australian/New Zealand Standards, Australia, 2004)

Australian/New Zealand Standards, *ISO 31000:2009 Risk Management – Principles and Guidelines* (Australian/New Zealand Standards, Australia, 2009)

B. Barret, The Garmin hack was a warning, *WIRED* (2020). Available online: https://www.wired.com/story/garmin-ransomware-hack-warning/. Accessed 29 Aug 2020

L. Bilge, T. Dumitras, Before we knew it: An empirical study of zero-day attacks in the real world, *ACM Conference on Computer and Communications Security*. North Carolina, USA, 2012: ACM

A. Cohen, Cyber (in)security decision-making dynamics when moving out of your comfort zone. Cyber Def. Rev. (Army Cyber Institute) **2**(1) (Winter), 45–60 (2017)

B. Engel, *Why CISOs Fail: The Missing Link in Security Management—and How to Fix It* (Taylor & Francis Group, London, 2018)

D. Gjerdrum, W. Salen, The new ERM Gold Standard: ISO 31000:2009. Prof. Saf. **55**(8), 43–44 (2010)

S.O. Hansson, Risk: Objective or subjective, facts or values. J. Risk Res. **13**(2), 231–238 (2010)

ISACA, State of cybersecurity: Implications for 2016, *RSA Conference*. San Francisco, CA, 2016: ISACA

D. Kahneman, *Thinking, Fast and Slow* (Penguin Books, London, 2011)

L. Kroll, Developing the connected world of 2018 and beyond, *Forbes*. (2018). Available online: https://www.forbes.com/sites/forbestechcouncil/2018/03/16/developing-the-connected-world-of-2018-and-beyond/#4d4c4f241e51. Accessed 19 Feb 2019

A. Madureira, F. den Hartog, H. Bouwman, N. Baken, Empirical validation of Metcalfe's law: How Internet usage patterns have changed over time. Inf. Econ. Policy **25**, 246–256 (2013)

R. Metcalfe, Metcalfe's law: A network becomes more valuable as it reaches more users. Infoworld **17** (1995)

G. Murray, M. Johnstone, C. Valli, The convergence of IT and OT in critical infrastructure, *The Proceedings of 15th Australian Information Security Management Conference*. Edith Cowan University, Perth, 2017: Edith Cowan University

National Audit Office, *Investigation: WannaCry Cyber Attack and the NHS* (National Audit Office, London, 2018)

National Institute of Standards and Technology, *NIST Special Publication 800-37 - Risk Management Framework for Information Systems and Organizations Revision 2* (U.S. Department of Commerce, Gaithersburg, 2018)

L.H. Newman, The biggest cybersecurity disasters of 2017 so far, *WIRED*. (2017). Available online: https://www.wired.com/story/2017-biggest-hacks-so-far/. Accessed 16 June 2018

L.H. Newman, The leaked NSA spy tool that hacked the world, *WIRED*. (2018). Available online: https://www.wired.com/story/eternalblue-leaked-nsa-spy-tool-hacked-world/. Accessed 6 June 2018

G. Nott, Visa changes "hurt us directly" says Atlassian's Cannon-Brookes, *CIO*, 13 March 2018 (2018). Available online: https://www.cio.com.au/article/634601/visa-changes-hurt-us-directly-says-atlassian-cannon-brookes/. Accessed 14 Mar 2018

B. Popken, Equifax execs resign; Security Head, Mauldin, was music major, *NBC News*. (2017). Available online: https://www.nbcnews.com/business/consumer/equifax-executives-step-down-scrutiny-intensifies-credit-bureaus-n801706. Accessed 24 Mar 2018

S. Scheferman, Ransomware predictions past, present, future, *ITSP Magazine*. (2016). Available online: https://itspmagazine.com/from-the-newsroom/ransomware-predictions-past-present-future-past. Accessed 20 May 2018

B. Strom, A. Applebaum, K. Nickels, A. Pennington, C. Thomas, MITRE ATT&CK: Design and philosophy, (2018). Available online: https://attack.mitre.org/docs/ATTACK_Design_and_Philosophy_March_2020.pdf. Accessed 12 Dec 2018

The University of New South Wales, Medical Studies/Doctor of Medicine, (2018). Available online: https://www.futurestudents.unsw.edu.au/degreetool/medicine/medical-studiesdoctor-medicine. Accessed 24 May 2018

R. Van der Meulen, *Gartner Says 8.4 Billion Connected "Things" Will Be in Use in 2017, Up 31 Percent From 2016*, 7 Feb 2017

M. Warner, Cybersecurity: Unis and TAFEs can fill the gap, *The Australian*. 27 Feb 2018 (2018) [Online]. Available online: https://www.theaustralian.com.au/business/technology/cybersecurity-unis-and-tafes-can-fill-the-gap/news-story/27551e3844d8a5269ed32d6eb7e7f4b0. Accessed 22 Apr 2018

Chapter 7
Auxiliary Impacts

Parallel developments in encryption technologies and cryptocurrencies has given rise to ransomware evolving into a prodigious cyberthreat. The impacts from this emergence are diverse, triggering a multitude of downstream effects on governments and organisations. Encryption is at the heart of cybersecurity, but its application can inhibit law enforcement processes and practices. It has also been the fundamental building block for ransomwares development. The rise of ransomware challenges the level of surety in organisational cyber risk management practices. Deciphering the outcomes of this parallel emergence is essential to further understanding the threat ransomware poses. The final section also provides an opportunity to reflect on the quandary ransomware creates between professional service providers and academia.

7.1 The Impact of Encryption Technologies on Law Enforcement

Analysis of global cybercrimes statistics indicates that historically, the most prevalent cybercrimes are "are identity theft, the theft of sensitive personal information, fraud, money laundering, and cyber-attacks for political or economic gain" (Treverton et al. 2011). Ransomware combines the underpinning techniques used in many of these cybercrimes, and from a data collection standpoint, it has generally been categorised as a type of cyberattack. Not dissimilar to most Internet-based businesses, the cybercrime environment is fluid, and organised cybercriminals can adapt rapidly to their environment and external market forces. For organised cybercrime syndicates, major law enforcement operations can be an incredibly disruptive force. For example, during periods where law enforcement agencies target

M. Ryan, *Ransomware Revolution: The Rise of a Prodigious Cyber Threat*,
Advances in Information Security 85, https://doi.org/10.1007/978-3-030-66583-8_7

ransomware proponents, cybercriminals adapt and may switch to alternative cybercrimes such as deploying crypto mining malware.[1]

Emerging technologies create endless challenges for law enforcement agencies. As new technologies emerge, cybercriminals rapidly seek methods to adopt and exploit them for their own financial gain. Many cybercrimes are traditional crimes that have evolved to use the Internet and other emerging technologies to aid and enable new criminal operations. From a law enforcement perspective, ransomware attacks can be considered a cyber-enabled version of extortion and kidnapping – instead of taking people or objects hostage, cybercriminals deny their victims access to their own data until a ransom has been paid (and potentially after).

The modern world is dynamic and so are cybercriminal organisations. Comparatively, law enforcement agencies are relatively static and generally subject to extensive bureaucratic processes. Law enforcement is traditionally considered to be a reactive process, with laws frequently being formulated or adapted to close loopholes after they have already been exploited by criminals. The process required to create or amend legislation to prevent or prosecute cybercrimes can also significantly range in complexity and duration between different countries and jurisdictions. The establishment and ongoing advancement of cyber legislation is further complicated by terminology nuances and incompetent politicians who have limited business and cyber acumen. Quite simply, the process of drafting and implementing new laws to prevent cybercrime is made more arduous if those responsible for the process do not understand how the Internet or emerging technologies work.

Cybercriminals are acutely aware of these long-standing political and legal problems, with cybercriminals commonly launching attacks and diverting funds across multiple geographic regions. The process of attack and payment concealment not only delays investigations, but it increases the complexity and resources required to investigate cybercrimes. This notion can be demonstrated by applying it to the process of investigating ransomware ransom payments. The investigation process may require the tracking of thousands of small payments across multiple currencies (both fiat and virtual) that have passed through potentially hundreds of different legal jurisdictions multiple times each. For the attacker, this complex laundering process can be easily created and automated using Application Process Interfaces (API). For law enforcement, tracing these transactions requires manual processing. The speed of technology, specialist user knowledge and level of anonymity of the Internet combine to inhibit law enforcement's ability to investigate, attribute and prosecute these types of crimes to a degree never experienced before.

Globally, many law enforcement agencies have made improvements to their capacity to detect and respond and their abilities to undertake cyber investigations. Despite these ongoing advancements in law enforcement, many countries continue to struggle with the pace of technology and criminal development. This struggle is not limited to developing nations, with cybercrime continuing to outpace law

[1] Note: Cybercriminals may also adapt of their own volition when the price of cryptocurrencies rise, simply because it provides a better return on investment for their efforts.

enforcement in developed nations. The lack of resources directly impacts law enforcement and intelligence workforces. Within the United States, the NSA is reliant on the advice and assistance from private specialists operating under contract with the government. It is estimated by early this decade, more than half of the US intelligence budget was spent on external contractors and that the number of these contractors accounted for around a quarter of the total intelligence workforce. The continued shift to externally based resources may further complicate the future of cyber law enforcement.

The Internet has emerged as a burgeoning, decentralised, multi-stakeholder environment with limited or no international institutional governance. Control and regulation of the Internet and its enabled technologies are considered a divisive topic in international relations today. There is no commonly agreed definition for cybercrime, and there is no overarching international cyber governance body. Despite the lack of international consensus, collaborative law enforcement efforts to disrupt, degrade, deter and prosecute organised transnational cybercrime syndicates have produced encouraging results.

The speed of technology in conjunction with corruption and blurred legal lines encourages cybercriminals to undertake ransomware attacks. These problems are exacerbated in many regions by law enforcement agencies being constrained to outdated investigative and legislative powers and tools. Enabling law enforcement agencies to protect us from ransomware attacks requires international collaboration and for governments to provide their law enforcement agencies with new technologies, legislative authorities and the financial resources required.

Whilst these collaborative law enforcement operations have been moderately successful, they are repressed by stringent oversight and compliance obligations with domestic and international laws. From a law enforcement perspective, Treverton et al. explain that "committing crimes across borders complicates the tasks of law enforcement agencies that are trying to combat them and can give the criminals flexibility in adapting their methods to counter or neutralize law enforcement initiatives" (Treverton et al. 2011). This complication of cybercrime legislation and enforcement is supported by cybercrime expert Tatiana Tropina (2013) whose research found that "organised criminal groups in cyberspace, both traditional ones and those operating solely online, remain – and probably will continue to remain – several steps ahead of legislators and law enforcement agencies" (Tropina 2013, p. 56). This reactive law enforcement process highlights the continuous speed of technology adoption by organised cybercrime syndicates.

Another consistent problem with cyberattacks is the lack of deterrence. Cyberattackers frequently operate and target victims on opposing sides of political disputes. This provides a layer of protection for their operations and significantly reduces the risk of apprehension and prosecution. There is also no shortage of allegations of cyberattacks such as ransomware attacks being state sponsored. Mezzour et al.'s (2014) research into the origins of cyberattacks identified "that many countries in Eastern Europe and a few countries in Central America are particularly attractive for hosting attacking computers. This is because these countries have a combination of good computing infrastructure and high levels of corruption"

(Mezzour et al. 2014). Corrupt officials provide multiple layers of protection to organised cybercrime syndicates through providing privileged information and safe haven and deliberately slowing or inhibiting investigations.

7.2 Ransomware's Building Block

The correlation of encryption with ransomware is tantamount but, on the surface, represents an elementary analysis. Deeper analysis indicates that encryption is at the crux of almost all technologies that enabled the rise of ransomware. The propagation of encryption has enabled and encouraged the development of a series of technologies that underpin ransomware attacks such as those detailed below in Fig. 7.1. Whilst many of these emerging technologies were not developed to advance cyber-attacks, they have been rapidly adapted and adopted by attackers to enhance their operations and delay attack attribution.

There is a bitter irony in the detail that modern encryption was designed to provide security for data at rest and during transmission; now, it represents an

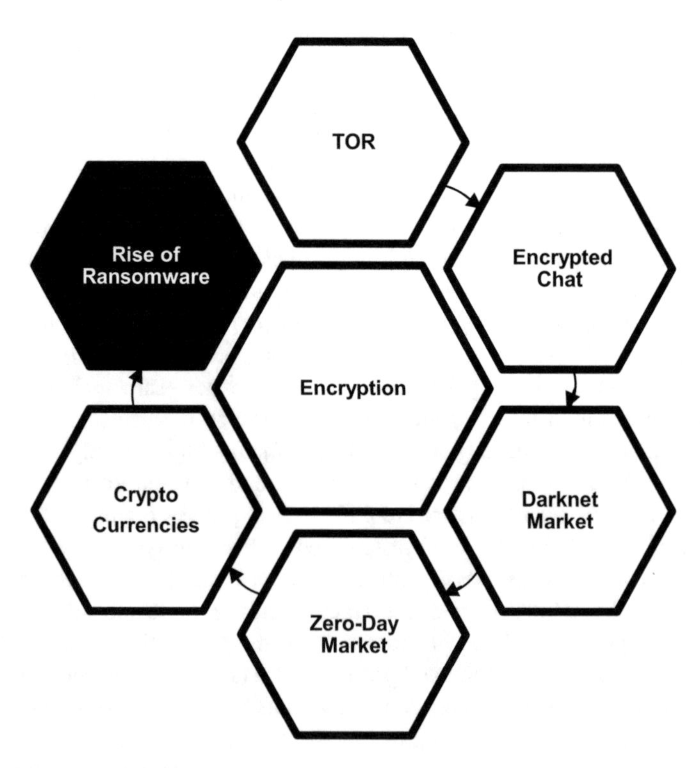

Fig. 7.1 Ransomware's building block

insurmountable hurdle for ransomware victims.[2] The devasting impact of cyberattacks has captured the attention of nations, who aggressively seek to develop and acquire zero-day exploits for future wars. In the public domain, the number of technology developers and number of devices in operation grow every day. Encryption is the building block that underpins the development of connected devices, which also suggest that the worst may be yet to come.

7.3 Applied Cryptography

Applied cryptography (encryption) is at the heart of today's financial and military communication and storage systems. The same algorithms that are used to protect our sensitive data at rest and during transmission are fundamental components of ransomware attacks. Encryptions functionality in ransomware attacks is more than just pure mathematics; it is an interdisciplinary domain that involves elements of many disciplines of science, engineering and humanities. The application of encryption is the fundamental fabric behind what enables the Internet to be anonymous. It is the foundational component of encrypted chat rooms, applications, virtual black markets, untraceable drop boxes, cryptocurrencies and super-secure operating systems that even the NSA are supposedly unable to crack (Levine 2017).

WikiLeaks founder Julian Assange argues that the universe believes in encryption because it is easier to encrypt information than it is to decrypt it (Assange 2012). Ultimately, this is because of the existence of a number of one-way functions. One-way functions are easy to compute in one direction, but given an output of the function, it is extremely difficult (meaning, in computational terms, time-consuming) to reconstitute the input (Joque 2018). This process, specifically when repeated multiple times using different algorithms, gives the attacker an almost insurmountable computational advantage. Whilst encryption inevitably becomes the hurdle that most ransomware victims are unable to overcome, it highlights one component of the attack. It is a logical argument that a primary reason behind the emphatic rise in ransomware attacks is the simplicity and availability of advanced encryption tools and techniques. There have been limited advances in mainstream encryption techniques in the past two decades. However, from an encryption application standpoint, the entry barriers to deploy advanced encryption techniques have significantly reduced throughout the same period.

Users of free (open-source) Internet-based platforms such as WhatsApp, LINE, Signal and Facebook messenger can easily apply AES 256-bit encryption to their everyday data communications. Multiple mobile phone and IoT device manufacturers

[2] Note: There is substantial evidence to indicate that when encryption is applied correctly, it is unsurmountable in the practical medium to long term by all existing computational models. Whilst it can be argued that encryption used in ransomware attacks has previously been deciphered, this was the result of a flawed malware design process, not a flaw or weakness in the encryption algorithm.

now offer inbuilt encryption software applications. There is also no shortage of advanced open-source Internet-based encryption applications. Quite simply, there has never been a point in time where it has been easier for someone with limited or no knowledge of encryption to deploy or apply advanced encryption techniques. Developers of ransomware attacks do not need to be mathematicians to deploy advanced encryption; they can simply download toolkits for free or purchase customised toolkits from other cybercriminals to develop and launch new ransomware attacks.

7.3.1 Encryption Backdoors

Encryption regulation has become a Gordian knot for security researchers, politicians, law enforcement agencies, privacy advocates and media outlets. Presently, there is a strong argument why law enforcement agencies through legitimate legal processes should be given access to user's stored encrypted data, and this is detailed below in Table 7.1 (Castro and McQuinn 2016, p. 13). At the same time, there is a strong argument from security researcher and privacy advocates that providing encryption keys or the insertion of backdoors weakens security. There is also widespread concern that these processes may enable abusive and illegal practices beyond their proposed uses. It has also been argued that unless universally applied, this would have limited or negative effects on security. According to a Harvard University study, "two-thirds of the nearly nine hundred hardware and software products that incorporate encryption have been built outside the United States" (Schneier et al. 2016). Castro and McQuinn present the argument in Table 7.1 below:

Irrespective of these arguments, the result will have no effect on the volume or complexity of ransomware attacks. The encryption algorithms commonly used in ransomware attacks are already open source and publicly available. Law enforcement, managed service providers, security researchers and victims already have unrestricted access to these encryption algorithms. What they do not have access to is the encryption keys, which are generated on a random, ad hoc or case-by-case basis. In all successful ransomware attacks, these keys will remain private (known only to the attacker) until released by the attacker or until an alternative method can be developed to recover the key.

Table 7.1 How encryption affects government access

	Data at rest	Data in motion
Law enforcement	Law enforcement is unable to access the encrypted data stored on the user's device or in the cloud (even with a valid warrant)	Law enforcement cannot use wiretaps to intercept communications
Intelligence agency	Intelligence agency is unable to access the encrypted data stored on the user's device or in the cloud, including bulk access to user data	Intelligence agency is unable to analyse communications for trigger terms

7.3.2 Potential Impacts of Quantum Computing

The race to harness quantum computing is considered by many experts to be a technological arms race. Whilst there will be spoils for the victor, the entire gains may be short-lived. One of the most significant gains that may potentially arise from quantum computing is the ability to rapidly decrypt intercepted encrypted messages and to breach (gain unauthorised electronic access to) secure networks and databases. For a nation state such as the United States, who is rumoured to be storing digital mountains of intercepted encrypted data, the breakthrough could be considered more significant than breaking the Enigma code.

The problem is that overtime, it can naturally be expected that other advanced major powers will be in a similar position to develop and deploy advanced decryption capabilities. As with all advanced military technologies, their true power is only realised once they have reached and triumphed on the battlefield. Today, the virtual battlefield is everywhere, and as a result, the technology will gradually be rolled out to more and more locations until it is eventually compromised. At this point in time (or perceived point), even without being compromised, governments, adversaries and criminal organisations will naturally modify their behaviour and modus operandi. Based on empirical evidence, this in turn will lead to more advanced types of encryption being developed, and the game of cat and mouse will begin the next round.

Quantum computing would have significant tactical advantages when trying to deploy or defeat AI, ML, or data analytics-based intrusion systems. This may allow the attacker to gain access to a foreign system through the application of previously unknown methods or manipulation of previously unknown vectors; however, at this stage, it is impossible to definitively identify all possible gains and impacts from quantum computing.

7.3.3 Anonymity on the Internet

Since the creation of the Internet, user security and privacy from virtual prying eyes has always been a priority for Internet users. To enhance Internet communication security, in the mid-1990s, the US Naval Research Laboratory established the Tor project, which aimed to develop an effective and secure method of communications which could protect the identities and locations of US intelligence agents operating in the field (Syverson 2016). The project was successful and later deployed as an open-source software (free to use) that was available to the general public in 2004. In the period since, Tor has attracted a large following of Internet users who wish to cloak their identities, geographical location and online activities.

Tor is not the only anonymous distributed network, but it is the most well-known and widely cited. Tor's users are diverse and include general citizens, state agents, whistle-blowers, journalists, political activists, cybercriminals, narco-traffickers and terrorists.[3] Tor was originally designed to as method to protect a user's identity whilst communicating online. Tor's ability to provide increased anonymity has attracted a large volume of Internet users, from both criminal and legitimate, who wish to operate behind a cloak of anonymity. In 2016, the then US Assistant Attorney General Leslie Caldwell explained "the protections offered by Tor shield illegal activities in parts of the internet dubbed the Dark Web; a marketplace used by criminals to sell drugs, weapons, dangerous toxins and child pornography" (Caldwell 2015). The term "Dark Web" (alternatively known as the Darknet) refers to the anonymous areas of the Internet. These areas are not to be confused with the Deep Web or Deepnet, which refers to websites that are hosted on the common Internet but are not indexed by search engines (such as Google and Yahoo). To access sites on the Dark Web securely and privately, users generally also use a dedicated Dark Web Internet browser such as Tor, I2P, Freenet or Disconnect.

Tor is a free encryption-centric software platform that is designed to protect the identity of its users by obscuring traffic analysis and degrading the ability of unauthorised parties to undertake network surveillance (Dingledine and Syverson 2004). Due to its level of anonymity, Tor is widely used as a secure method of underground communications. The system is architecturally described as:

> Onion routing is a distributed overlay network designed to anonymize TCP-based applications like web browsing, secure shell, and instant messaging. Clients choose a path through the network and build a circuit, in which each node (or "onion router" or "OR") in the path knows its predecessor and successor, but no other nodes in the circuit. Traffic flows down the circuit in fixed-size cells, which are unwrapped by a symmetric key at each node (like the layers of an onion) and relayed downstream…Perfect forward secrecy: in the original onion routing design, a single hostile node could record traffic and later compromise successive nodes in the circuit and force them to decrypt it. Rather than using a single multiply encrypted data structure (an onion) to lay each circuit, Tor now uses an incremental or telescoping path-building design, where the initiator negotiates session keys with each successive hop in the circuit. Once these keys are deleted, subsequently compromised nodes cannot decrypt old traffic (Dingledine and Syverson 2004).

A simplified version of how Tor's network architecture functions is also illustrated below in Fig. 7.2 (Tor Project 2018):

Whilst Tor is considered to provide anonymous communications, Geelkerken has argued that the level of Internet anonymity can be further improved by *Alice* simultaneously acting as a Tor node. This change in configuration adds an additional layer of complexity because it is difficult to determine which data is produced by *Alice* and which is the throughput of other Tor users (Geelkerken 2006). This change is detailed in Fig. 7.3:

Cybercriminals and legitimate Internet users may also benefit from the use of a Virtual Private Network (VPN). A VPN creates an encrypted connection (tunnel) over the Internet between a user device and a network. This tunnel ensures that data (packets) are transmitted safely by concealing the true source of the packet. If

[3] Note: Due to the nature of the activites undertaken on Tor, state agents, law enforcement and intelligence researcher are key users of the platform.

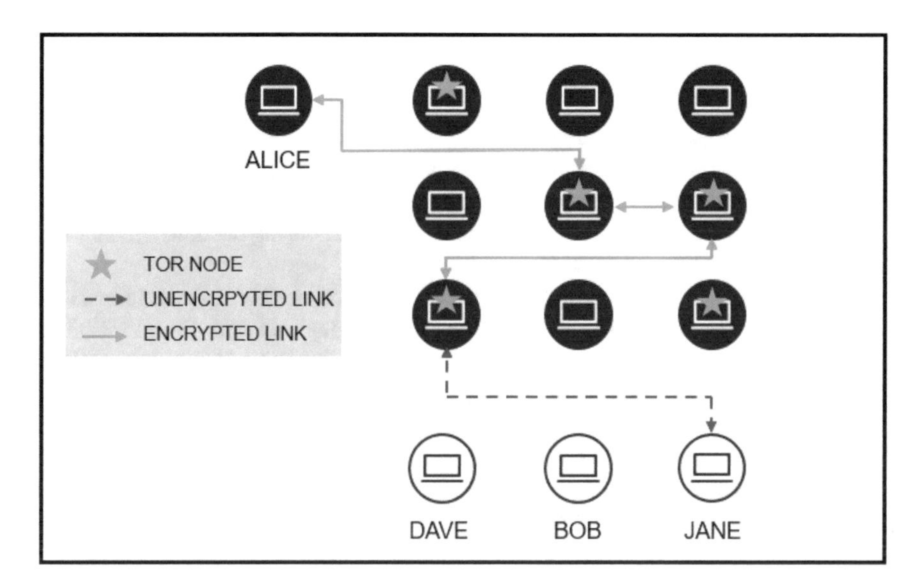

Fig. 7.2 How Tor works simplified

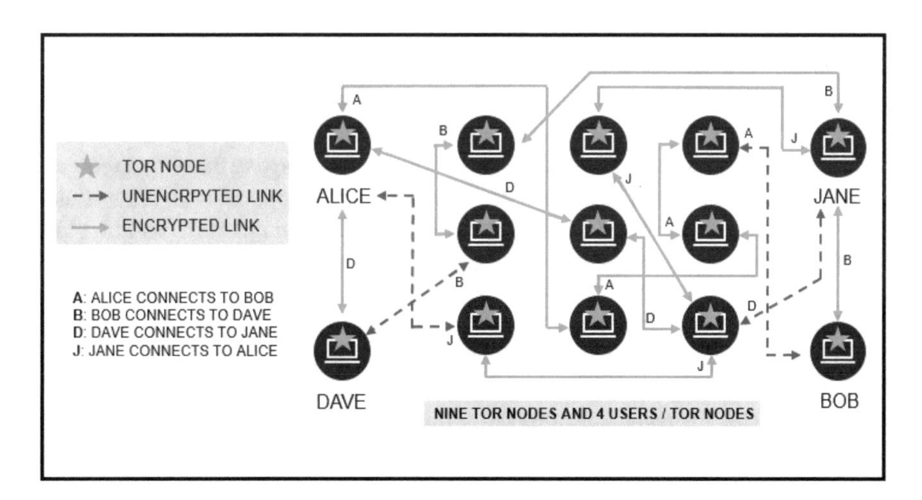

Fig. 7.3 Alice as Tor node

configured correctly, this concealment may provide an effective security measure against eavesdropping. VPNs were originally developed to allow users to remote into networks to undertake work (i.e., work from home). They are also commonly used by Internet users to circumvent Internet content restrictions. The combination of using Tor and a VPN may restrict an attacker's bandwidth; however, this may have limited impact on ransomware attacks due to their high level of script automation, and the attacker ultimately utilising the victim's own device to encrypt the data.

This ability to communicate and operate anonymously on the Internet is crucial for the development and deployment of ransomware attacks. Encrypted platforms such as Tor and VPNs enable cybercriminals to discuss criminal activities in forums and private messages with limited or no threat of eavesdropping by law enforcement agencies. These types of encrypted platforms allow organised cybercriminals and organisations syndicates to operate transnationally in real time whilst actively reducing their inherent risk of discovery and attribution during the development and planning phases of ransomware attacks.[4]

Additionally, these evolving encrypted platforms enable cybercriminals to remotely launch ransomware attacks from multiple physical and virtual locations concurrently. Analysis of major ransomware attacks indicates that cyberattackers often commandeer innocent Internet users' networks and devices in order to distribute and launch ransomware attacks. This process can significantly impact the resources available to the attacker. This process can also inhibit or prevent the attribution of ransomware attacks by creating layers of protection for the attacker. All these processes seek to further reduce the risk to the attacker, and structurally, they mirror the classic approach to security of defence-in-depth.

7.4 Considerations for Practice

Even without the ransomware threat, organisations already need to constantly re-evaluate their preparedness to effectively manage the changing cyberthreat landscape. This research demonstrates that ransomware is an adaptive threat that requires constant additional demanding defensive measures and additional mitigation strategies. Defeating ransomware may demand not just additional measures that fit into a broad cybersecurity strategy at corporate level but a stand-alone plan and dedicated staff resources. From a risk management approach, many cyberthreats can be grouped together because they share similar countermeasures (controls) or because the anticipated impact of an incident is below an acceptable financial or operational threshold. This grouping of cyberthreats is a common industry practice that can enhance the risk management process by improving the visibility of major cyberthreats at the executive level. The grouping of cyberthreats can also provide great insight into what organisations perceive as their most significant cyberthreats or exposures.

For instance, within large financial institutions, the loss of funds through fraudulent or theft-related cyber activities is considered to be a constant threat. Social engineering attacks such as phishing campaigns present a serious threat and can occur every second. Distributed Denial-of-Service (DDoS) attacks against IT infrastructure are a significant cyberthreat that have the ability to rapidly disrupt and

[4] Note: It can be argued the advent of cryptocurrencies was the final piece of the puzzle required to maximise ransomware's criminal potential.

deny customers and staff access to critical Web-based services. The corruption of data, whether accidental or malicious, presents another significant cyberthreat for financial institutions. Whilst these examples are all significant cyberthreats, this research hypothesises that ransomware now stands alone atop of these cyberthreats.

This view is shaped by the premise that ransomware exhibits an ability to combine multiple elements from other cyberattack forms; it can propagate through phishing emails, corrupt user data and deny users access to systems and services, all whilst inducing crippling losses to the organisation through lost productivity and extortion demands. This prodigious power is further exacerbated by ransomware's potential attack longevity. Typically, cyberattacks are a significant encumbrance for victims; however, their impacts are typically short term. An inherent problem with ransomware is even with a mature cyber defence capability, the time required to recover critical systems may be days, weeks or even months. Within the United States, some healthcare providers were still trying to fully recover their systems from a recent ransomware attack 2 months after they refused to pay a ransom (McGee 2019).

There is no single measure for preventing ransomware attacks. Therefore, preparation and containment must be at the core of mitigating the potential threat. Denying the existence or complaining about the cost or complexity of the problem will do nothing to reduce the risk. As organisations continue to rapidly adopt emerging technologies, their attack surface will continue to expand, which in turn increases the size and complexity of perimeter security. Understanding that ransomware attacks will continue to occur despite robust policies, processes and defenders' best efforts is crucial to managing the risk. Acknowledging this predicament reinforces the common conclusion among scholars and practitioners that containment should be at the crux of any ransomware mitigation strategy, and containment in this case implies enhanced segmentation of data and system components.

Such containment in large enterprises can create complex architecture problems as resilience against most cyberthreats is commonly enhanced and achieved through interconnectedness. Improvements in automation technologies have substantially contributed to the enhancement of resilience through connectedness within large organisations. At the same time, it is this high degree of interconnectedness in many cases that enables ransomware to self-propagate and spread rapidly across an organisation. The challenge of simultaneously achieving containment and resilience based on connectedness can be further complicated by ageing and diverse architecture standards and infrastructure, which is commonly referred to as technical debt.[5]

[5] Note: Technical debt is common within large enterprises who have undertaken a series of minor upgrades which has created environmental complexity. The term originally referred to the amount of extra programming work that would be required to continue to operate software over a given period. However, the term now broadly refers to the level of unnecessary complexity through lots of incremental changes to a system, which can be exacerbated by work being undertaken by several people who might not fully understand the original design.

Successfully achieving containment should be tested through training exercises that encompass people, processes and technology across the entire organisation.[6] Developing and running ransomware training exercises is no easy feat for large corporations and government agencies. However, from a practical perspective, if an organisation can't afford the time to effectively execute training exercises, then how can it afford to incur a real ransomware attack? Training exercises do not guarantee success against thwarting ransomware attacks, but they can provide a repeatable metric for organisational preparedness.

Whilst enterprise risk management practices are contrary to many cybercriminal practices, much can be learned from the business practices and organisational structures used by organised cybercrime syndicates. The adaptive nature of cybercrime requires streamlined flexible business processes and agile organisational structures that can minimise losses from potential windows of opportunity for criminal operations. The rapid adoption and deployment of emerging technologies is not without risk, and cybercrime syndicates need to constantly evaluate their situation to avoid detection, interdiction from law enforcement activities and the activities of other organised cybercrime syndicates.

Global ransomware attacks since 2016 have confirmed ransomware as a prodigious cyberthreat. The speed of impact, coercive force, attack effectiveness, duration and cost of remediation provide stark indicators why organisations should be proactively taking action to prevent and prepare for ransomware attacks. This rise in threat should have also triggered changes across cybersecurity business practices, altering people's behaviour, processes and how technologies are managed across all levels of the organisation.

7.4.1 Risk Profiles Must Change

This research argues that ransomware is the by-product of numerous encryption technologies emerging and being deployed in unison. The deployment of these technologies and the success of ransomware have forced corporations and governments to revise their cyber risk profiles. With a growing number of high-profile ransomware victims, executives are compelled to buy down their own organisation's risk. This is a resource-demanding process that ultimately impacts the organisation's bottom line. With no definitive set of countermeasures, ransomware challenges senior executives to manage known multifaceted cyber risks whilst still providing maximum return on investments for stakeholders.

This is a complex challenge to manage because ransomware by design is a dynamic, not static, threat. Managing dynamic threats requires dynamic people,

[6] Note: Incident response plans and exercises should also incorporate third-party suppliers and providers. Most organisations are now reliant on multiple third-party service providers to commonly provide services such as cloud storage and hosting, identity and access, authentication, human resources and payment platforms.

Fig. 7.4 Risk profile

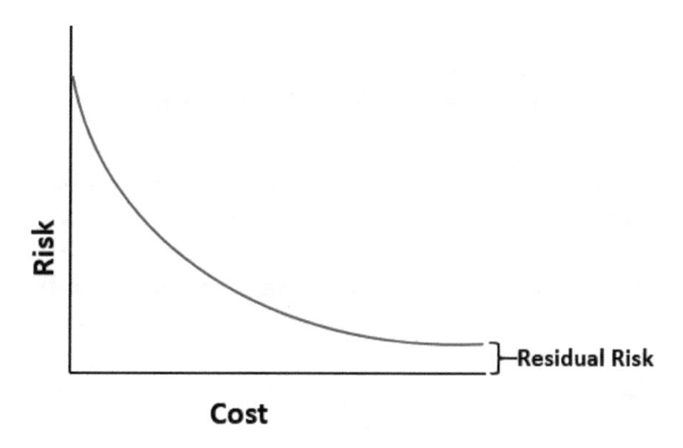

processes and security controls. This can be problematic for large organisations due to the cost and time requirements needed to roll out enterprise-wide changes. Figure 7.4 below illustrates that organisations are compelled to seek an effective balance between the cost of security and risk exposure. It also highlights that irrespective of security expenditure, there will always be a level of residual cyber risk.

To effectively minimise and manage the risk associated with ransomware, many organisations will require behavioural changes in addition to financial investment. Security should be at the heart of all decisions that relate to business technologies, customer data and technology infrastructure. As corporate boards and elected officials begin to enter an era of increased cybersecurity scrutiny and accountability, the ability to make practical and cost-effective cybersecurity decisions pertaining to ransomware defences will be increasingly valuable. Since 2013, ransomware has continued to demonstrate a unique ability to rapidly expose those who do not respect or understand it and even those who do but are unprepared for it.

7.5 Reflections on a Professional Services and Academic Quandary

Cybersecurity is an emerging academic field. The origins of the field began as an applied extension to the field of computer science; however, the field is an amalgamation of numerous sub-disciplines which draw from the fields of IT, business, warfare, risk management, law, engineering and behavioural sciences, to list but a few. As noted in the literature review, one of the better efforts to describe a "science of cybersecurity" that can be distinguished from its constitutive disciplines is that it is concerned mainly with the "organisation of defences" in a sociotechnical sense and not with the design of attack packages which is largely (though not exclusively) a question of computer science or engineering. The academic field is rapidly evolving with universities scrambling to provide established professionals advanced degrees in cybersecurity. The push for universities to provide more advanced degrees can be linked to a changing business attitude that now considers cybersecurity to be an everyone, not just an IT department, problem. The focal point for many of these new programs has been existing professionals looking to upskill or transfer fields, not undergraduates (Cabaj et al. 2018).

Whilst universities were relatively slow in developing cybersecurity programs, industry organisations such as Information Systems Audit and Control Association (ISACA), System Admin, Audit, Network, Security (SANS) and International Information System Security Certification Consortium (ISC^2) developed a series of professional certifications designed for the professional services and cybersecurity industry. The certification process generally requires a combination of practical experience in conjunction with a formal examination process. This approach has been successful, with many professional services and corporations preferring industry-certified professionals over university degrees (ISACA 2017). This creates a downstream effect as students don't require a degree to enter or advance their careers in the industry. Therefore, the next generation of security professionals may never start or return to university for further education in the field of cybersecurity.

Instead, many students and existing professionals enter the cybersecurity job market through direct or alternative methods, opting for industry certifications to advance their careers. This decision is made easier by expensive university fees and extreme shortages in the professional cyber skills market. As a result, many cybersecurity professionals did not and may never attend university; this also adversely impacts the quality and volume of academic research being produced by universities. These opposing approaches create a disconnect, which further extends the divide between academic institutions and the professional cybersecurity services industry.

This disconnect may be further demonstrated by the "Big 4" consultancies dropping the requirement for consultants to have a university degree (Singhal 2017). Whilst there may be a combination of reasons that initiated this change, it cannot be ignored that the same consultancies have aggressively expanded their cybersecurity service offerings and workforces throughout the past two decades. The reputation of these firms is a natural lure to attract top cyber talent, and this creates a concentration of cyber professionals in organisations that are either unable to or have limited interest in interacting with universities.[7]

These elements create a quandary between professional service providers and academic institutions. With reduced access to industry experts and events, the academic field has struggled to produce research that is relevant and engages with corporations and business leaders. With limited insight into corporate cybersecurity operations or cyber risk management practices, academia continues to produce research that has limited practical applications. It is easy for a researcher to say "if an organisation implements these seven security controls, they can prevent and

[7] Note: Cybersecurity experts employed in the professional services industries are typically prohibited from discussing major cyberattacks due to confidential agreements with their clients and employers. Simultaneously, the preferred skill benchmarks are industry certifications. This encourages major professional service providers to collaborate and focus their recruitment efforts on industry certifiers, not universities.

respond to ransomware attacks". The problem is that an organisation may have thousands of systems and millions of devices that cannot simply be shut down to in order to roll out the proposed control changes. It is also possible that a new attack vector may appear the next day, rendering the proposed controls uplift mute.

Not dissimilar to nation states, large corporations are locked in a continuous cycle of competition. This cycle has led to the rapid deployment of emerging technologies and services with limited or no security controls. The process of managing and uplifting these security controls is complex and requires specialist people, planning and resources, and there is a global deficiency of all three. To improve the level of engagement with the cybersecurity industry, academic institutions will need to demonstrate confidentiality and the ability to bring value to the professional service providers and their clients.[8]

References

J. Assange, *Cypherpunks: Freedom and the Future of the Internet* (OR Books, New York/London, 2012)

K. Cabaj, D. Domingos, Z. Kotulski, A. Respício, Cybersecurity education: Evolution of the discipline and analysis of master programs. Comput. Secur. **75**, 24–35 (2018)

L. Caldwell, The threat today, in *Cybersecurity + Law Enforcement: The Cutting Edge*, (U.S. Department of Justice, Bristol, 2015)

Castro, D., A. McQuinn, Unlocking Encryption: Information Security and the Rule of Law, *Information Technology and Innovation Foundation*. (Mar 2016)

N.M.R. Dingledine, P. Syverson, Tor: The second-generation onion router, *13th Conference on USENIX Security Symposium*. San Diego, 2004: USENIX

F. Geelkerken, TOR: The Onion Router. (2006). Available online: https://www.iusmentis.com/society/privacy/remailers/onionrouting/. Accessed 22 Feb 2018

ISACA, *Survey: Cyber Security Skills Gap Leaves 1 in 4 Organizations Exposed for Six Months or Longer*, San Francisco, CA, 13 Feb 2017

J. Joque, Distributed denial of service: Cybernetic sovereignty, in *Deconstruction Machines*, (University of Minnesota Press, Minnesota, 2018), pp. 111–148

Y. Levine, The Crypto-Keepers: How the politics-by-app hustle conquered all. The Baffler **36**(Fall), 66–79 (2017)

M. McGee, Cleaning up after ransomware attacks isn't easy, *Information Security Media Group*. (2019). Available online: https://www.careersinfosecurity.com/cleaning-up-after-ransomware-attacks-isnt-easy-a-12921. Accessed 20 Aug 2019

G. Mezzour, L.R. Carley, K. Carley, Global mapping of cyber attacks, *School of Computer Science Carnegie Mellon University*. (2014). Available online: http://casos.cs.cmu.edu/publications/papers/CMU-ISR-14-111.pdf. Accessed 23 Sept 2018

B. Schneier, K. Seidel, S. Vijayakumar, A worldwide survey of encryption products, 1, (2016). Available online: https://www.schneier.com/academic/paperfiles/worldwide-survey-of-encryption-products.pdf. Accessed 12 Jan 2019

[8] Note: It can be argued that cybersecurity uncertainty is in the financial interest of vendors and professional service providers. Whilst these vendors and consultancies endeavour to provide advice and products that protect their clients, security attacks and uncertainty directly benefit the profitability of their businesses.

P. Singhal, PwC to end university degree employment requirement, *Sydney Morning Herald*. Sydney, 30 April 2017 (2017) [Online]. Available online: https://www.smh.com.au/education/pwc-to-end-university-degree-employment-requirement-20170424-gvrb7c.html. Accessed 29 Aug 2017

P. Syverson, Brief history: Onion routing. (2016). Available online: https://www.onionrouter.net/History.html. Accessed 19 Mar 2019

Tor Project, Tor: Overview. (2018). Available online: https://www.torproject.org/about/overview.html.en. Accessed 20 Mar 2019

G. Treverton, M. Wollman, E. Wilke, D. Lai, The threat will continue to morph, in *Moving Toward the Future of Policing*, (RAND Corporation, 2011), pp. 89–106

T. Tropina, Organized crime in cyberspace, in *Transnational Organized Crime*, ed. by H. Böll-Stiftung, R. Schönenberg, (Transcript Verlag, 2013). Freiburg, Germany

Chapter 8
Failed Translations

When examining why ransomware has been so successful as a form of cyberattack, a logical investigative process requires the researcher to scrutinise popular configurations of cyber defences designed to prevent ransomware attacks. This chapter explores how the classical strategy of defence-in-depth has inadvertently influenced the spread of ransomware. There is also usefulness in examining the contrast in applying defence-in-depth strategies and principles in physical versus the cybersecurity domain. Although it is not plausible or practical to examine the role individual security controls or products play in the prevention of each particular strain of ransomware, there is value in analysing the adoption of underlying cyber strategies and principles that government and private enterprises have commonly adopted and adapted to design their own cyber defence systems.

The formulation and design of modern enterprise cyber defences have generally been rooted to a series of adopted strategies and principles from alternative time periods and disciplines. Typically, these strategies reference some form of connection or obscure relationship to military warfare, the defence of medieval castles, or that they were even translated directly from Sun Tzu's *The Art of War* (Sun Tzu 1971). Whilst all of these strategies, philosophies and defensive tactics may have been effective in their respective eras, this section argues that their adaption into modern cyber defences has been relatively unsuccessful in stopping ransomware attacks. Indeed, I am not the first academic or security practitioner to raise the failed premise and problems associated with adopting these foregone strategies (Wolff 2015; Kewley and Lowry 2001; Prescott 2012). This section explores whether this failure is the result of failed philosophical translations and strategy incompatibility or whether it is simply the result of the ever-increasing complexity in design and implementation of cyber defences.

M. Ryan, *Ransomware Revolution: The Rise of a Prodigious Cyber Threat*,
Advances in Information Security 85, https://doi.org/10.1007/978-3-030-66583-8_8

8.1 Defence-in-Depth

Within the cybersecurity industry, the term defence-in-depth commonly refers to a security strategy whereby security controls are deployed in a series of connected layers to achieve cyber defence. The term defence-in-depth origins can be traced to Beeler (1956) who coined the term to describe a series of castles which he observed to have been positioned in a strategic defensive formation to protect London at the turn of the first century (Beeler 1956). Beeler observed that a series of castles were strategically positioned on the two most geographically accommodating fronts leading to London in a manner that created a defence-in-depth arrangement. This created a scenario whereby an attacking force would be lured into a series of battles where the defenders would be able to repeatedly counter-attack from positions of strength. As castles on the perimeters began to be overrun, the remaining defenders would fall back to the next castle, taking up another position of strength for the next round of the battle. The strategy was designed so that the defenders were consistently able to counter-attack from superior battlefield positions, which enabled the defenders to repel attackers with superior numerical forces for a greater period of time.

The other primary objective of the defence-in-depth strategy is to prolong the battle duration. By prolonging the battle duration, defenders were provided time to call on, prepare and position reinforcements for the coming battles. This strategy also has the additional advantage of potentially reducing the strength of the attacking force through attrition. Whilst an attacker wounded during the heat of battle may have been able to continue to fight to the end of the current battle, the extension of the conflict by a day or week may have resulted in the warrior's health detreating, which potentially removed him from the forthcoming battles.

Whilst there were clear tactical advantages to adopting a defence-in-depth strategy to defend London, the underlying reason for its deployment was actually a cost-saving strategy. Adoption of the strategy enabled the number of guards in each castle's garrison to be reduced, which in turn allowed the monarchy to maintain the existing level of perimeter coverage with fewer guards. This reduction meant that London could now be defended with a reduced number of perimeter guards, thus lowering the ongoing security overhead required. Alternatively, should the total number of guards deployed remained constant, it could have been argued the strategy was primarily adopted to increase perimeter coverage (enhanced detection capability); unfortunately, that was not the driving force in this scenario.

8.1.1 Translating Defence-in-Depth

Inspired by the military strategy of deploying multiple controls in a series of layers, defence-in-depth is a common best practice recommendation amongst majority of cybersecurity consultancies and vendors. Despite originally being developed as a

Fig. 8.1 Perceived defence-in-depth principles

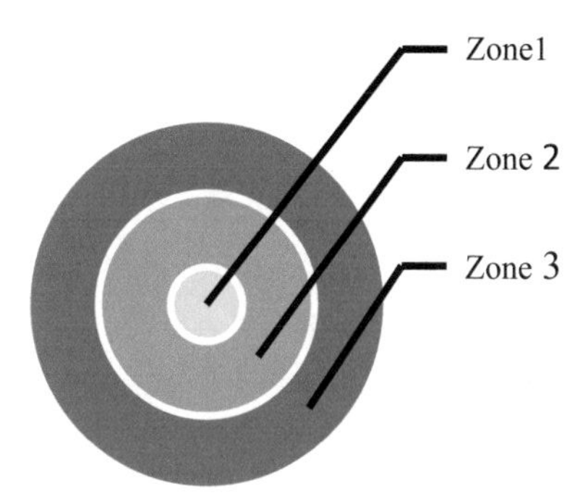

mechanism to buy additional time to organise defences against invading armies, cybersecurity practitioners have commonly translated this layering to literally mean an increased level of security assurance. The following illustration depicts how the principles of defence-in-depth have been commonly translated to depict some form of cross-cut set of Babushka dolls (Fig. 8.1).

Whilst the origins of defence-in-depth have little or no correlation with the cybersecurity of enterprises, this has not deterred security practitioners, vendors and enterprises themselves from adapting and adopting defence-in-depth as their underpinning strategy for designing and deploying their cyber defences (controls). Analysis of empirical cyber defence strategies reveals that defence-in-depth was adopted by enterprises well before the risks were even known because we have been using this strategy to protect computer systems longer than the existence of modern systems and today's cyberthreats.[1] This notion is supported by Wolff who observed that:

> The idea that computer defences should be combined and layered to reinforce each other is often invoked under the umbrella notion of defence-in-depth a term so vaguely defined and inconsistently applied in computer security…that as a guiding philosophy this is both untrue and unhelpful (Wolff 2015, p. 19).

Figure 8.2 highlights the translation being used to create a subsequent adaption model for an alternative security design purpose. The common theme depicts security being achieved or enhanced through additional layers being implemented into the design. Kewley and Lowry (2001) argue that traditional cybersecurity "thinking lends itself to the philosophy that the more layers of protection you add, the more secure a system will be… In the cyber domain, however, multiple layers of defence do not necessarily add together to create a higher level of assurance" (Kewley and Lowry 2001). These errors in translation can be consistently detected in the cybersecurity discourse from both a philosophical and its practical implementation. The mainstream adoption of these failed translations is both problematic and potentially dangerous because these translations are commonly used to formulate the underlying

[1] Note: Defence-in-depth most likely was adopted from physical security. It would have been a familiar term due to its frequent use in the doctrine related to the physical security of assets for governments agencies and enterprises.

Fig. 8.2 Example of defence-in-depth adaption for cyber defence

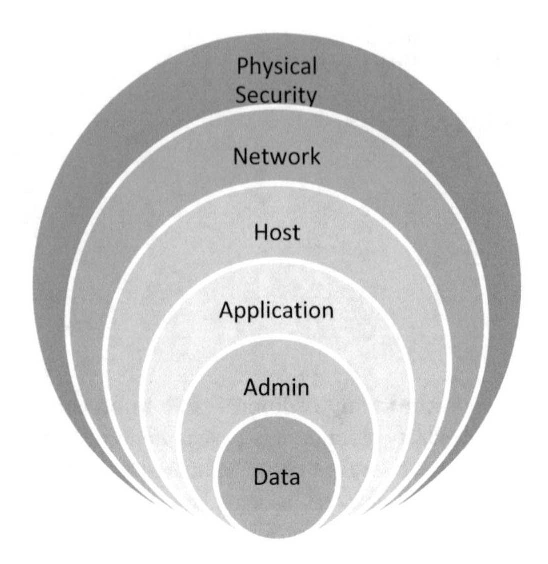

strategy that determines which, where and how cybersecurity controls should be deployed in enterprises to defend against ransomware attacks.

8.1.2 Defence-in-Breadth

When examining defence-in-depth, it is important to distinguish the differences between defence-in-depth (control layering) as a strategy for designing and deploying cyber defences such as firewalls and Intrusion Detection Systems (IDS) versus deploying a strategy of defence-in-breadth (control abundance). Prescott (2012) argues that the best outcome security practitioners can hope for is an array of security products and services that are highly effective which can complement each other. The concept is that what is missed by some will hopefully be detected by another – hence the term defence-in-breadth. The fundamental difference is cybersecurity controls are deployed and designed in a manner which aims to be complementary with each other, not a series of overlapping layers (Prescott 2012).

It is also essential to understand that the ability to recover data from multiple repositories is not a strategy of defence-in-depth but indeed a strategy of defence-in-breadth. From a practical perspective, defence-in-breadth is an expensive and complex cyber defence strategy as it can be problematic to design, test and replicate with any high degree of assurance. Fundamentally, the concept of defence-in-breadth accepts that breaches of cybersecurity will occur; however, by design, it is reliant on controls abundance to detect these breaches. This is not an ideal strategy for designing cyber defences against ransomware attacks because it does not provide control assurance by design (i.e., controls monitoring other controls). However, it may be suitable strategy for designing ransomware and other associated business continuity recovery strategies.

8.2 Evolution of Warfare

For thousands of years, military strategists have devised plans to defend their lands in the event an army of invaders appeared at their shores. These strategies have had to constantly evolve because warfare has always pursued the capability to strike an enemy at an increasing distance whilst trying to remain protected. This evolution is present throughout history and can be observed in the deployment of archers, catapults, guns, aircraft, missiles, strategic stand-off weapons, the use of Unmanned Autonomous Vehicles (UAVs) and in more recent times cyberattacks. As attackers developed improved abilities to strike targets at increasing distances, defences have had to evolve to mitigate the rising threat. The continued correlation between the warfare evolution, archaic security strategies and cybersecurity is idiocy. To put it simply, the majority of these approaches are considered archaic because they were all defeated emphatically hundreds of years ago.

8.2.1 Fort Eben-Emael

The discussion of enterprise cyber defences being designed and somehow aligned to defending castles by the construction of a moat is misguided, but nonetheless, it is an intriguing idea that should be briefly explored. In simplistic terms, the purpose of a moat is to buy the defender time (by creating a physical obstacle to overcome) whilst simultaneously preventing items (i.e., ladders or structures) from easily being placed against the castle walls. However, the underlying objective of the moat is not only to delay the attacker but to simultaneously place the attacker in a vulnerable position where they can be killed or injured. From an empirical perspective, it can be argued that this simple approach for defending castles remained relatively effective for hundreds of years.

Transposing this idea of castles and moats into more modern times rekindled my memories of studying the 1940 German assault on the Belgian Fort Eben-Emael for a special operations planning course. The Fort Eben-Emael is located on the border of Belgium, Luxembourg, and the Netherlands and was of strategic importance for the German advance through Europe. The fort was a hardened facility located on the edge of a canal creating a moat around the perimeter on the German side. The bridges leading to the territories surrounding the fort on the German approach were also rigged for destruction to prevent the advance of a large assaulting force. The fort was aided by natural terrain obstacles and were purposely designed and constructed to withstand aircraft and artillery attacks (Fig. 8.3).

The Fort Eben-Emael applied defence-in-depth principles to mitigate all conceivable types of German attack. Despite all the strategies and fortification efforts, the Germans were able to defeat a fortified force twice their numerical size and capture the fort in just over a day whilst incurring only 44 casualties (Kuhn 1978). The Germans were able to achieve this outcome by orchestrating an attack that used

Fig. 8.3 Fort Eben-Emael (Kliem 1940)

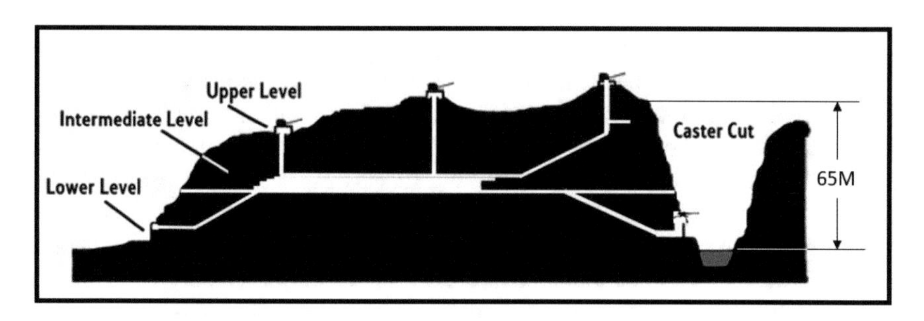

Fig. 8.4 Cross section of Fort Eben-Emael

gliders to enable a silent long-range insertion. The attack was launched in early hours of the morning, enabling the German soldiers to land at the fort and other strategic locations at sunrise. This method of entry concealed their insertion and provided them the element of surprise, affording them relative superiority (McRaven n.d.), which paved the way for a successful attack (Fig. 8.4).

Whilst the battle of Fort Eben-Emael has limited correlation with modern enterprise cyber defences, it does highlight that even the most well-known modern cyber defence-in-depth analogy was defeated almost a century ago. This historic tale also raises the strategic and defensive differences between conflict and crime and the actors which operate in each and both environments. Whilst state actors have increasingly been linked to cybercrimes, this has not translated into cyber defenders being afforded the permission to counter-attack. Acknowledging this inability to counter-attack is fundamental to understanding the limitations of adopting or adapting a cyber defence strategy that is built on the concept of defence-in-depth.

8.3 States Versus Enterprise Cyber Defences

Advanced military strategies and technologies have always been a source of envy for commercial enterprises, but not all military strategies to defend assets are translatable to commercial enterprises. As an example, the use of lightweight chain-link fences to defend the perimeter of military bases is common practice around the world. Whilst effective at establishing a defined physical barrier that deters and prevents unauthorised entry, the fences themself offer little resistance against a skilled or determined attacker. Nevertheless, in most scenarios, these humble chain-link fences are generally fit for their intended purpose because they are not being used in isolation to defend high-value assets. However, just behind this thin wire metal façade lies the real deterrent.

The primary reasons people do not commonly break into military bases is two-fold: (a) Firstly, they don't need anything from within the base, so there is limited value to gain from breaking into these types of facilities versus other commercial sites, and (b) you never know when a group of soldiers with machines gun are going to appear behind you. That uncertainty is pivotal to the underlying success of the defence strategy, which is deterrence. States achieve deterrence because the perceived reward is generally not worth the potential risk of being discovered. Whilst private enterprises can achieve deterrence relatively easily in the physical domain,[2] this is an extremely difficult proposition to achieve in the cyber domain.

The fundamental principle of defence-in-depth is reliant on the defender's ability to counter-attack. Without the ability to counter-attack, defenders generate no direct threat or element of deterrence toward the attacker. As a result, the cyberattacks keep coming because the attacker's relative force strength is not decreasing or diminished in any way. Their cost may be increasing, which is reducing their potential ROI, but they are not losing men or equipment in the ongoing battle. In this scenario, the attack will continue until the defenders can isolate their network from the attacker (segregate, deflect or shut down), the attacker withdraws or, in the case of an automated attack, an impasse is reached. Without the ability to counter-attack, the vast majority of enterprises will be unable to successfully execute a defence-in-depth strategy in the cyber domain. However, this is not an argument for enabling enterprises to become offensive cyber warriors; it is the result of analysing the effectiveness of the defence-in-depth-derived strategies against ransomware attacks.

From a practical perspective, this outcome indicates that most defence-in-depth cyber defences should be categorised as a form of cyber defence-by-deflection, not defence-in-depth. To fundamentally be considered an effective defence-in-depth strategy, there is an underlying requirement to counter-attack. This failure or inability to legally counter-attack also raises the hypothesis that defence-in-depth-derived

[2] Note: Enterprises generally create deterrence policies in the guise of codes of conduct to deter criminal activities such as fraud and theft. This is reinforced through the deployment of security controls such as monitoring software, security guards, CCTV and the promotion of acceptable social standards.

cyber defence strategies, such as those deployed by state agencies and armed forces, may not be suitable for enterprises because they lack a fundamental requirement of the strategy. The following section examines the differences between defence-in-depth in the cyber and physical domains and how deterrence can be achieved in the physical domain by private enterprises.

8.4 Cyber Dreams of Physical

Amongst cybersecurity practitioners, there is a common belief that they inherently understand the nuances of physical security. We have all heard the stories. Three years ago, they attended Defcon and picked a couple of off-the-shelf padlocks – master locksmith achievement unlocked. Discipline completed right? By contrast, most physical security practitioners do not pretend or claim to understand cybersecurity. There is widespread belief in the security industry that achieving physical security is a much simpler task to achieve; however, I would argue this is not necessarily a reflection of difficulty but the nature of physical security being more intuitive. Ultimately from a professional perspective, they are two different sub-disciplines that require different approaches and skill sets, and I have encountered few security practitioners who have sufficient expertise to be a professional in both.

When cybersecurity practitioners design physical security, it is analogous of hiking up Mount Kilimanjaro, tripping over a rock and dislocating your finger. You turn to your climbing group to ask if anyone is a medical doctor. Your request falls silent; however, moments later, a veterinarian steps forward to kindly offer you their expertise. Considering your predicament, this may be your best option to solve the problem at hand. However, in an ordinary scenario, a veterinarian would not be your desired professional to treat the problem.

From a design perspective, physical security defence-in-depth strategies require multiple security layers, which should be increasing in robustness and/or complexity compared to the previous outer layer. The following series of scenarios and diagrams illustrate why defence-in-depth can be a sound strategy for physical security whilst highlighting potential limitations of the strategy's adaptation in the cyber domain. In each scenario, the security controls have been designed and deployed in layers so as the attacker advances closer towards to the crown jewels, the security controls should increase in robustness and/or complexity to defeat, to slow the attacker's progress (Fig. 8.5).

When designing security for retail banks, there are multiple security events that the bank will be designed for and prepared to encounter. In my experience, three of the most common attack vectors that are anticipated in the security design of a retail bank branches are:

(a) Armed robbery[3]

[3] Note: In the United States, during 2018, almost 95 percent of bank robberies occur at the bank tellers (cashier) counter. See Federal Bureau of Investigation (2018). Available online: https://www.fbi.gov/file-repository/bank-crime-statistics-2018.pdf/view [Accessed 23 May 2020].

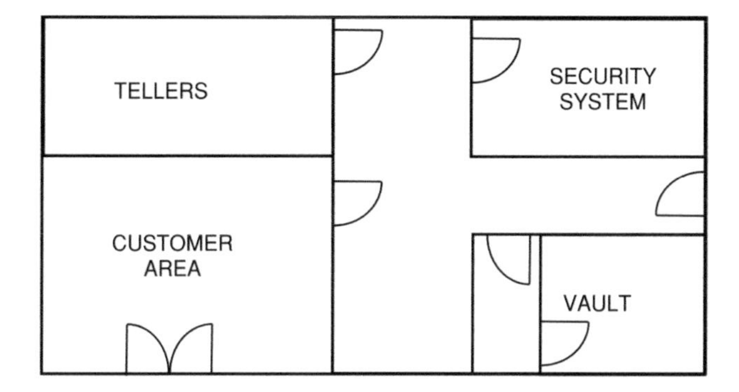

Fig. 8.5 Example of bank layout

(b) Burglary (after hours)
(c) Hostage scenario (staff or customer)

It is also worth noting that not all bank security incidents are related to the theft of money. Retail banks may become the site for security events that range from a political campaign (i.e., terrorism) to the escalation of a domestic dispute involving a customer or staff member.

8.4.1 The Armed Robber

Armoured robberies are challenging scenarios for banks to completely defend against. Whilst an armed attacker may be able to quickly obtain cash from tellers (cashier), the total cash being held by tellers continues to reduce. This is a security mitigation, but it is also the result of society transitioning towards a cashless environment. At the outset of this attack, even as the attacker prepares to enter the bank, the first control they will encounter is CCTV cameras.[4] There will be numerous cameras deployed at intersecting angles which cannot be easily reached; therefore, to defeat this security control, an attacker may elect to wear a disguise. This will conceal their identity, but it will also alert everyone inside and potentially coming into the bank that a robbery is occurring. Detection of the attacker will trigger a silent alarm, and the clock is now running. The attacker advances and receives cash from the teller. At this stage, the attacker has two relatively straightforward options: (a) leave the bank with the small amount of acquired cash or (b) continue to advance

[4]Note: Commercial grade CCTV cameras have internal detectors, so spray painting or damaging the camera is generally not an effective method to defeat them, and it will most likely just trigger the alarm system or alert the guard monitoring the branch.

the attack towards the vault. Should the robber advance the attack, they will need to gain access to the staff area and proceed to the vault.

By this stage, the vault is already locked down (linked to alarm system), and the alarm will increase the time delay required to open the vault and, in some scenarios, may have even been designed to prohibit the vault from being opened without external permission (remote interlock system). These basic security controls have bought the defenders time, and the police are now on site. Every path forward from this point will be a complex minefield that challenges the attackers' risk versus reward calculations, drastically increasing the attackers' odds of being caught or killed.

8.4.2 Burglary

Excellent choice, however, this method of attack will most likely require more than one person, so the attacker will need to assemble a crew before they can begin the attack. From a point of entry, the bank's rear door is as good of a choice as any place to launch the attack. Whether the attackers decide to attempt to pick the multiple locks or opt to simply cut through (or sledgehammer) the door down, the attackers have already tripped multiple security controls before they got through the door, and the clock is now running. How is this possible? Well, as the attacker quickly finds out, these are not the locks they picked at Defcon, and they also have sensors in them that detect the attempted breach. The doors and walls are also fitted with vibration sensors and reed switches, but realistically, none of these security controls were actually needed because your presence was already detected by the CCTV cameras looking at the bank's rear door. The detection process was assisted because modern CCTV cameras have movement detectors that are linked to the security alarm system, which alerted the guard in the security monitoring centre who notified the authorities. Using this attack methodology, an attacker may make it through the first door, but they are going to require significantly more time to open or breach the vault, and that is something now beyond their control.

Alternatively, maybe the attacker tries to remove some bricks from the outer wall or tries coming through the ceiling or floor. The problem is all of these potential entry points have security controls that have detected the attacker's attempted intrusion and notified the police, and the attacker is still no closer to opening the vault yet. Worse yet, the attacker may even become trapped in the roof or a wall cavity, thus being unable to escape when the police arrive. Ultimately, in most cases, the attackers' endgame is to open the vault. This is also problematic because vaults have evolved and, for decades now, have been designed to defend their contents from not only robbers but also fires, floods and large, angry mobs (i.e., times of civil unrest). The evolution of their design has also made them highly resistive to explosive and vehicle-orientated attacks. It is becoming clear that this attack vector is not working, maybe it is time for a different approach.

8.4.3 Hostage

The final choice in this brief foray into bank robbing is taking a hostage. The attacker elects to take the manager hostage, maybe you could take him or her the night before or grab them on their way to work in the morning. This was a common tactic from previous eras, and as a result of its prior success, most banks today now require two or more staff to be present when the bank is opened. They also require two or more to be present to open the bank's vault. Most modern banks have also transitioned away from rear access and now only allow staff to enter through the front door. This could be problematic because banks are generally located in positions of moderate to high pedestrian traffic (the bank was situated there for a reason). This may complicate attacker's plan but does not prevent it, so they elect to move forward with the attack.

As the attacker continues with the attack, they are able to gain access to the bank by threatening both staff, but they still have to contend with the cameras, security system and the vault. Once inside the bank, there are numerous security measures throughout the bank that may detect the attacker's presence, and detection by any of these will prevent the vault from being opened. Banks understand the security and safety risk posed by these types of attacks on their staff, so they have implemented elaborate security controls and processes to discretely detect these types of attacks. Some of these advancements include staff not being able to access the security system. This could be as simple as physically placing the security system beyond the access of the branches staff or relocating it to a remote location; these measures increase the complexity and knowledge required to defeat the system. In fact, most bank staff are unaware of the security controls in place because they do not have a need to know. Ultimately, their job is to serve the bank and its customers, not to design or reconfigure the branches security system.

8.4.4 External Reference Data

The purpose of this section is to provide insight into bank robberies and to apply an additional point of reference. In the United States during 2018, the bank's alarm system was triggered in over 90 percent of armed robberies, with CCTV capturing over 95 percent of robbery-related incidents (Federal Bureau of Investigation 2018). In Australia, it is expected that this statistic could be much higher. This is because Australian banks are designed using significantly more advanced security systems. This design difference is the result of significantly lower labour in the United States, which has created a security model that prefers low-cost guards instead of expensive security systems. In the coming years, I would anticipate that globally, more banks will deploy expensive advanced security systems as labour costs continue to rise and due to cloud-hosted security systems enabling large banking enterprises to service a large number of retail and commercial sites using one system.

8.4.5 *Scenario Results*

Whilst applying only some of the security controls available, it is clear through each scenario that defence-in-depth is an effective security strategy for defending bank branches.[5] In the bank scenario, security controls can easily be deployed in layers that overlap each other, which prevents key security controls from being defeated without an alternative security control detecting the attacker or attempted attack. This could be as simple as a vibration sensor mounted to a door, which is linked through the security system to a camera that can observe the door. The camera watching the door is in turn visible by another camera, and both are monitored by a guard on-site or remotely. There are no shortages of examples that could be used to highlight the effectiveness of defence-in-depth in the physical security domain. But how does that impact cybersecurity? The answer to that question is a paradox because it is both a lot and very little.

8.5 Diverging Translations

One clear and distinguishing feature that can be observed when comparing physical security and cybersecurity adaptions of defence-in-depth is security vendor and practitioners' interpretations of security control layers. As depicted earlier, both physical and cybersecurity layers are commonly depicted as butting up against each other without any gaps, whereas in practical deployments, they are not actually side by side. In cybersecurity designs, there are virtual gaps which are commonly not eliminated, whereas in advanced physical security designs, the security controls not only butt up against each other, but they actually overlap each other.

Detailed analysis of physical security designs reveals that these overlapping controls are configured to be more than a series of controls which are simply connected to the same network to report their current status. Instead, they operate as a series of sensors and controls working in unison to form an interconnected system. In multiple scenarios, this configuration not only enables an individual control to report when they are coming under attack, but another control in the system is. Furthermore, in some scenarios, this control overlapping design exhibits the ability to detect the attacker prior to the attack even commencing.

Additionally, because the controls are designed to operate as a closed loop system, this increases the overall attack complexity required to defeat the system. As a result of this increased system complexity, more time and resources are required to defeat the individual controls, which in turn increases the time staff, and ultimately, the police have to respond to the attack. Figure 8.6 below depicts how a series of controls are commonly configured to create an alarm system (electronic access con-

[5] Note: To maintain the ongoing integrity of bank branch security, numerous security controls and processes were omitted from these scenarios.

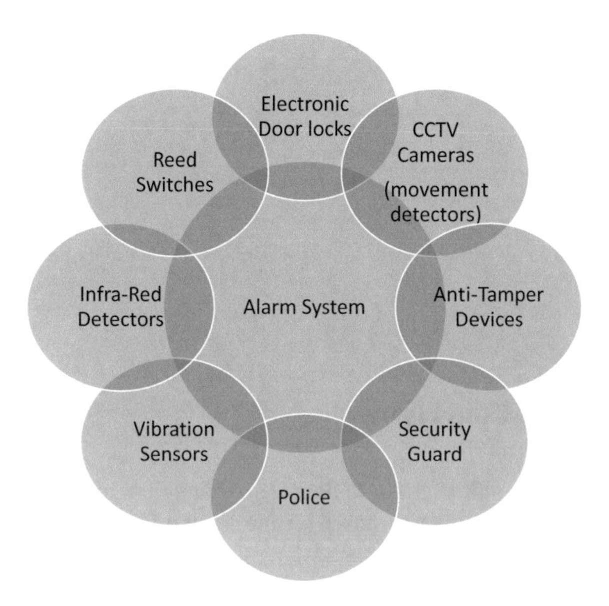

Fig. 8.6 Basic alarm system sensors

trol system). The crown jewel of an electronic alarm system is the head-end (server), which is why this system is generally located in a secure rack, within a secure room located behind multiple layers of overlapping controls.[6] Head-end systems should also be designed with a high degree of resilience to detect and maintain operations in the event of a power or telecommunications networks disruption.

In the retail bank scenario discussed above, the alarm system represented just one of the challenges that needed to be overcome to complete a successful attack reaching the vault. In practice, overcoming the alarm system is a complex challenge because the physical security design of the bank integrates multiple security controls across multiple domains to protect the vault's contents. The illustration below further highlights how security controls from multiple domains are designed to overlap each other (Fig. 8.7).

These simple physical security examples depict how it is relatively easy to design rudimentary systems to work in unison to create complexity, which can inhibit an attacker's ability to successfully execute an attack. This does not mean they cannot be overcome, but they do have a significant impact on the risk/reward ratio and the ROI needed to complete an attack.

---

[6] Note: Modern physical security systems can also be located off-site in another location such as the commercial headquarters or even a data centre. Remote-operated federated physical security systems have been common practice for critical infrastructure for decades. Beyond the potential cost advantage, federated systems also prevent local staff from making changes to their own security systems and enable security administrators to be frequently rotated and their system actions closely monitored.

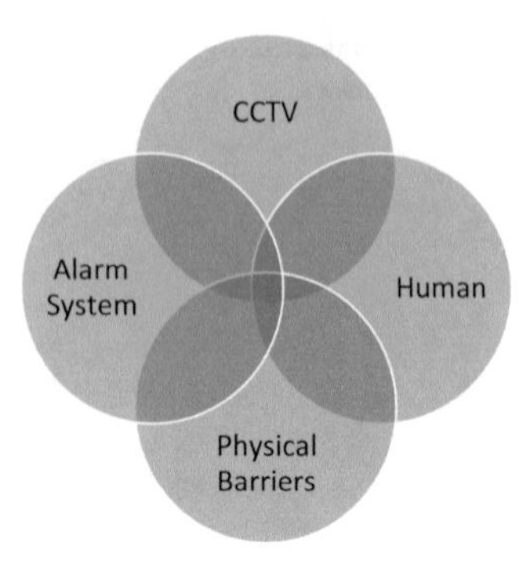

Fig. 8.7 Defence-in-depth synergy

8.6 Conclusions

Philosophically, the underlying strategy behind the physical security adaption of defence-in-depth is based on altering the risk versus reward profile for known attack vectors. As the attacker advances, the control robustness and complexity required to defeat the controls increase. Whilst this design cannot predict or repel every type of attack that may come in the future against a control, the adaption exhibits a strong ability to dictate the location of the battle required to defeat the control. By denying the attacker access to specific spaces, this complicates the attacker's ability to easily launch concealed low-risk (of detection) attacks. This is where the defence-in-depth translation diverges for cyber because this (virtual) defensive high ground cannot be easily or consistently achieved in the cybersecurity domain.

In practice, the cyber domain has limited or no ability to rapidly reveal a skilled attacker's true identity. Whilst advancing a ransomware attack, an attacker may inadvertently trip a cybersecurity control (detection tool); however, this does not significantly increase their chance of being identified. Tripping an alarm in the cybersecurity monitoring system whilst trying to encrypt data or steal funds may (or may not) increase the complexity of the attack required moving forward, but this does not necessarily start the timer until the police arrive at the attacker's door. In the case of ransomware, being detected may be an insignificant detail for an automated attack because this attempted attack may simply be chalked up as another statistic on a graph in a cybersecurity report depicting one of a million cyberattacks detected and defended by the enterprise for this month. This is also why is it essential that enterprises continue to focus on accurately measuring and reducing their "breakout"[7] and "dwell times."[8] Increasing the time it takes to laterally move whilst reducing the time an adversary can retain a foothold in a network is pivotal to preventing both automated and human-directed ransomware attacks.

[7] Note: The term "breakout time" is commonly described as the time it takes an attacker to laterally move to the next point in the network after gaining initial access.

[8] Note: The term "dwell time" is commonly described as the time an attacker can remain within a network after gaining initial access.

References

J. Beeler, Castles and Strategy in Norman and Early Angevin England. *Speculum* **31**, 581–601 (1956)

Federal Bureau of Investigation, *Bank Crime Statistics* (Federal Bureau of Investigation, Washington, D.C., 2018)

D. Kewley, J. Lowry, Observations on the effects of defense in depth on adversary behavior in cyber warfare, *IEEE Workshop on Information Assurance and Security*. United States Military Academy, West Point, NY, 2001

Kliem, Fort of Eben Emael, in B. Bundesarchiv_Bild_146-1971-011-29, _Fort_Eben_Emael,_ Albert_Kanal.jpg, Creative Commons, 1940

V. Kuhn, *German Paratroops in World War II* (Ian Allan, London, 1978)

W. McRaven, *The theory of special operations*. Master of Arts in National Security Affairs Naval Postgraduate School. (n.d.)

S. Prescott, Defense in depth: An impractical strategy for a cyber world, *SANS Institute*, (2012). Available online: https://www.sans.org/reading-room/whitepapers/warfare/paper/33896. Accessed 11 May 2020

Sun Tzu, *The Art of War*. S. Griffith. (Oxford University Press, New York, 1971)

J. Wolff, *Classes of Defense for Computer Systems*. Doctor of Philosophy in Engineering Systems: Technology, Management, and Policy Massachusetts Institute of Technology, June 2015

Chapter 9
The Sociology of Ransomware

The cybersecurity industry has been quick to adopt existing risk mitigation strategies and new cybersecurity products to defend against ransomware attacks. However, limited research has analysed whether there are any causal factors associated with ransomware attacks. Causal factors may range from why an individual develops or launches a ransomware attack to why ransomware attacks continue to be effective against large corporations. Are ransomware attacks so technically advanced that large corporations cannot stop them, or are enterprises failing to mitigate their risk effectively? Society understands that ransomware poses a serious threat, but we do not have a clear understanding of what factors may be contributing to its further development or exacerbating the problem.

9.1 The Sociology and Psychology of Ransomware

With limited research available that is specific to ransomware, this research draws on classical behavioural studies in sociology, psychology and criminology that are associated with other forms of cybercrime. Whilst the cyber aspect is an essential component of cybercrime, fundamentally, cybercrimes are not different to the same crimes that are undertaken in the physical world. However, it must be acknowledged from a heuristic approach that the virtual and perceived anonymity elements of cybercrimes may substantially alter the decision-making process of an individual contemplating committing a cybercrime. As a result, it could naturally be expected that this may significantly alter the probability of an individual undertaking a cybercrime due to the cyber element of the potential crime being perceived by the individual as an advantage or disadvantage. This assumption is based on the premise that financially motivated cyberattackers are rational actors.

M. Ryan, *Ransomware Revolution: The Rise of a Prodigious Cyber Threat*,
Advances in Information Security 85, https://doi.org/10.1007/978-3-030-66583-8_9

Cybercrime as a sub-discipline within the sociology, psychology and criminology discourses is currently in the infancy stage of its development. At present, there is an extremely limited body of academic research that has evaluated the specific psychological, sociological and criminological factors, influences and motivations that lead to or contribute to an individual undertaking a ransomware attack. There is also a deficiency in academic research that has analysed the demographic, geographical or socio-economic factors that influence cybercrime offences more broadly.

One of the most comprehensive bodies of work in the field to date is a research paper by Europol, Middlesex University and the UCD Geary Institute for Public Policy. This research collaboration by Aiken et al. (2016) draws on existing evidence about:

> Online behaviour and associations with criminal and antisocial behaviour amongst young people. Specifically, the research was designed to explore the trajectories and pathways that lead to 'cyber-criminality' through a series of mixed-methodological endeavours and the integration of theoretical frameworks across criminology and psychology, including cyber-psychology and computer science. The potential pathway from technology talented curious youth, to juvenile cyber-delinquent, to lone cybercriminal to organised cybercrime was considered (Aiken et al. 2016).

The research is supported by research from the Australian Bureau of Crime Statistics and Research in Harris (2015) that discovered "fraud offences committed by people under 18 years of age had jumped by 26 percent in the previous two years, and 84 percent in the previous three-year period" (Harris 2015). The rapid rise was in part attributed to Internet-based fraud and cybercrime; however, the research did not assess social factors or root causes behind the spike in cyber-related offences. It is also well established that throughout the formative teenage years, teenagers are more impulsive and more likely to engage in risk-taking behaviours.

Another difficulty with data cited from Australian cybersecurity research is reliability and the lack of transparency. Whilst the evidence indicates the number of fraud offences continues to climb in Australia, there is limited ability to correlate and categorise the incidents that are enabled or purely cyber-based. Currently, the Australian Bureau of Statistics (ABS), Australian Federal Police (AFP), NSW Police and the Australian Criminal Intelligence Commission (ACIC) provide no publicly available crime statistics tools or methodology to categorise crimes as a cybercrime. The inability to collect, categorise and accurately analyse cybercrime data comes at a time when Prof David Lacey (2017), the managing director of IDCARE, Australia and New Zealand's national identity and cyber support service, received over 28,000 requests for cyber assistance in 2016 (Lacey as cited in Wordsworth 2017). Without access to accurate and real-time data, cybersecurity research will continue to struggle to produce informative and meaningful results.

The rise of online black markets has also enabled the emergence of Crime-as-a-Service (CaaS) and RaaS services. These types of criminal services allow users who are not as tech-savvy to rent professionals to deploy advanced cybercrime techniques on their behalf. Looking to the future, this may lower the technical skills and experience required to develop attacks, thus allowing younger Internet users to easily undertake criminal activities without fully understanding the seriousness or repercussions of their crimes (Aiken et al. 2016). This changing virtual criminal environment is occurring during a period where "people are becoming more technically sophisticated; younger generations are using technology on a daily basis in school, learning digital technology at a very early age" (Ablon et al. 2014b, p. 35).

Another research study into economic crime by Coleman (1992) indicated "that economic crime is encouraged by the societal climate in which individualism is prevalent or promoted" (Coleman 1992). This supports Mars's (1982) earlier research into workplace economic crimes found that the "propensity for criminal behaviour increases when social relations become anonymous and short-lived" (Mars 1982). Whilst in 1982 Mars could not have predicted the impact the Internet would play in shaping today's society, his correlation of anonymity and short-lived relationships in economic crimes may be considered significantly more prevalent in theorising factors within modern cybercrimes.

9.2 Attacker Motives

Analysis of mainstream ransomware attacks to date indicate it is possible to derive primary motives for launching ransomware attacks. The first and most prevalent motive for ransomware attacks is financial gain. Cybercriminals and organised crime syndicates develop and initiate ransomware attacks as though they are legitimate business operations (Trend Micro 2018). This is supported by Symantec's *Internet Security Threat Report* which indicates the amount of money ransomware attacks generate keeps going up. And victims keep paying up despite contentious warnings from law enforcement and cybersecurity experts not to pay.[1] In 2016, the average ransomware attack made $1,077 (USD), an increase of 266 percent from the previous year. However, Symantec reported this reduced throughout 2017 to $522 (USD) as the result of criminal organisations switching their focus to illegally mining cryptocurrencies. Since the peaks in 2017, cryptocurrencies have significantly reduced in value, and this may drive cybercriminals to return to undertaking ransomware attacks to sustain their revenue streams.

The secondary motive is revenge, which is generally derived from an actual or perceived grievance. In 2016, *Business Insider* reported the story of a French security researcher who tricked cybercriminals into installing ransomware on their own systems after they infected his parents' computer with ransomware. The researcher convinced the attackers to accept a picture of his credit card, which contained an embedded piece of malware (Price 2016). This concept of active defence or striking back is not new, and robust debate continues on both sides of the argument. The active defence theory is supported by Rabkin and Rabkin who argue "the

[1] Note: Despite law enforcement advising citizens not to pay ransoms, there are no shortage of law enforcement and government organisations who have paid ransoms to recover their own systems. See Francescani (2016).

United States should let victims of computer attacks try to defend their data and their networks through counter hacking" (Rabkin and Rabkin 2016). Their research calls for private organisations to be permitted by US lawmakers to experiment with more aggressive tactics of active defence; however, they have also argued that we should be prepared to step back if that course turns out to be the more prudent course. Cyberattacks motivated by revenge are not constrained to hacking back; they also include attacks against innocent victims.

To date, there have been limited incidents of ex-employees or jilted lovers infecting their ex's computers with ransomware, but these types of incident should not come as a surprise in the future. In one incident in late 2015, Christopher Grupe was dismissed from Canadian Pacific Railways for insubordination and not playing well with others. Prior to returning his company laptop, he used it to corrupt the company's administration accounts, deleting essential files and events logs from the system to cover his tracks. System failures ensued shortly after his departure, and the company was forced to call in a third-party consulting firm to fix the network, who subsequently discovered and alerted the company to Grupe's indiscretions (Wood 2018).

Another common motive, particularly in teenagers, is intellectual curiosity. For many, being able to overcome the security of a protected network is perceived as a challenge to test their technical expertise. Whilst innocent at the outset, overtime, this has the potential to manifest into ego and superiority complexes, and there are no shortages of cybercriminals who have undertaken cyberattacks to showcase their skills and to show off. This line of motive was described by the not-for-profit hacking group Lulzsec, who famously proclaimed that many of their cyberattacks were "for the Lulz" (Norton 2011). Figure 9.1 below details another example of these types of ransomware attacks; in this scenario, victims of the ransomware attack were confronted with a request for payment seeking nude images, not financial payment.

Another known motive is to establish a smokescreen (Blaze 2017). In this scenario, a disgruntled employee or the business entity itself may install ransomware onto a network to conceal data such as event logs or financial transactions. This may be an effective measure to prevent the disclosure of sensitive or financial information or to prevent the discovery of evidence concerning the individual or organisation who may be being investigated by law enforcement.

9.3 Rational Choice Theory

The standard theoretical approach within law and economics for explaining criminal behaviour originates from the observation that potential criminals are rational decision makers (Becker 1969). Whilst rational choice theory is not the sole theory within sociology to hypothesise the causal motivations for criminal behaviour, it does have a strong empirical tradition in economics (Lindenberg 1992). McAdams and Ulen (2008) explain that:

Fig. 9.1 Example of non-financial-motivated ransomware (MalwareHunterTeam, 'Sample', Available online: https://twitter.com/malwrhunterteam/status/910952333084971008/photo/1)

> The theory assumes that potential criminals compare the expected costs and benefits of criminal activity, where the expected benefits include the anticipated monetary and non-monetary returns to the crime, discounted by their probabilities of realization, and the expected costs of the crime, which include formal and informal sanctions (the latter including loss of lawful employment opportunities, social stigma, and guilt), discounted by the probabilities of detection. If the expected benefits exceed the expected costs, then the rational potential criminal commits the crime; otherwise, he or she does not. Moreover, the rational potential criminal compares the expected costs and benefits of criminal activity with those of legitimate activity and rationally allocates her time and other resources between those alternatives so that the marginal net benefit is equated (McAdams and Ulen 2008, p. 2).

In Merton's (1968) classic analysis of American society in *Social Structure and Anomie,* money is a "crime-inducing character. It becomes manifested more precisely in situations in which money permeates the culture and people are 'bombarded' with images that stress the importance of financial success, while its power of attraction is strong enough to neutralize any ability of other values to influence or curtail the desire for it" (Merton 1968). The concept that money is beyond its physical trading value is echoed by Kahneman in *Thinking, Fast and Slow* who states that "money is a proxy for points on a scale of self-regard and achievement" (Kahneman 2011, p. 342).

Messner and Rosenfeld expanded on Merton's early works to reason that "when a value is attractive enough, the choice between the different means for attaining it is reduced from a moral to a purely technical problem. It thus becomes a question simply of how to most effectively attain the goal" (Messner and Rosenfeld 2001, p. 64). Analysis by Kharraz et al. specifically into ransomware attacks observed that "cybercriminals continuously strive to find more reliable charging methods by improving two important properties: (1) the difficulty of tracing the recipient of the payments, and (2) the ease of exchanging payments into a preferred currency" (Kharraz et al. 2015, p. 13). Engdahl's (2008) research challenges these theories by arguing that "a common feature in this line of thinking is its predominantly theoretical or abstract nature, and the fact that no empirical evidence is normally provided for verification purposes" (Engdahl 2008, p. 166). This argument challenges whether there are alternative theories that could be applied to further understand ransomware attacks.

9.4 Expected Utility Theory

Expected utility theory is a "process of how to choose rationally when you are not sure which outcome will result from your acts" (Bernoulli 1738). The theory provides a framework that can be applied to the decision-making process. For the example detailed in Table 9.1 below, the organisational decision to purchase or not to purchase cyber insurance, "each column corresponds to a state of the world; each row corresponds to an act; and each entry corresponds to the outcome that results when the act is performed in the state of the world" (Briggs 2017).

Applying Bernoulli's process in conjunction with Briggs's revised method "having set up the basic framework, the user can now rigorously define expected utility. The expected utility of an act *A* (for instance, purchasing cyber insurance) depends on two features of the problem" (Bernoulli 1738).

The following example by Briggs states "the value of each outcome - measured by a real number called a utility, and the probability of each outcome is conditional on *A*. Given these three pieces of information, *A*'s expected utility can be defined as (Briggs 2017):

$$EU(A) = \sum_{o \in O} P_A(o) U(o)$$

Where O is the set of outcomes, $P_A(o)$ is the probability of outcome o conditional on A, and $U(o)$ is the utility of o. The next two subsections will unpack the conditional probability

Table 9.1 Example of expected utility framework

		States	
		Breach	No breach
Acts	Buys cyber insurance	Financially encumbered, safe	Financially encumbered, safe
	Does not buy cyber insurance	Financially unencumbered, vulnerable	Financially unencumbered, safe

function P_A and the utility function U". The term $P_A(o)$ represents the probability of o given A — roughly, how likely it is that outcome o will occur, on the supposition that the agent chooses act A. The term $U(o)$ represents the utility of the outcome o — roughly, how valuable o is. Formally, U is a function that assigns a real number to each of the outcomes (Briggs 2017).

Expected utility theory has multiple applications in rational choice decision-making. The same process may be undertaken by an attacker contemplating launching a ransomware attack. The attack may be successful and generate income; however, it may also bring known consequences.

9.5 Prospect Theory

Within behavioural sciences, expected utility theory has dominated the analysis of decision-making under risk. Generally, it has been accepted as a normative model of rational choice. However, departing from classical rational choice theory, prospect theory analyses ubiquitous features of human thinking. Prospect theory is the work of economic Nobel laureate Daniel Kahneman and his late research partner Amos Tversky. The theory was originally developed to demonstrate the errors with rational choice theory and expected utility theory when applied to practical risk decision-making situations. Kahneman and Tversky (1979) state that:

> Many economic decisions involve transactions in which one pays money in exchange for a desirable prospect. Current decision theories analyse such problems as comparisons between the status quo and an alternative state which includes the acquired prospect minus its cost…The theory can also be extended to the typical situation of choice, where the probabilities of outcomes are not explicitly given. In such situations, decision weights must be attached to particular events rather than to stated probabilities, but they are expected to exhibit the essential properties that were ascribed to the weighting function (Kahneman and Tversky 1979, p. 288).

Camerer (1989) argues that "prospect theories may be serious contenders for replacing expected utility theory at least for specific purposes; in part because there is considerable empirical support for both reference-dependence and decision weights" (Camerer 1989). Prospect theory provides a tool to further analyse cybercrime and cyber risk management. From a behavioural science perspective, evaluating the intricacies of ransomware, the decision-making process of a criminal undertaking a ransomware attack is no more or no less important than the decision-making process of a victim deciding whether to pay the requested ransom. Both are gambling on risky outcomes that are mathematically predictable, yet the outcomes are uncertain. This is described as "decision making under risk and can be viewed as a choice between prospects or gambles. A prospect (x1, Pi; ...; xn, pn) is a contract that yields outcome xi with probability Pi, where Pl + P2 + ... + pn =1" (Kahneman and Tversky 1979, p. 263). Moore's research even indicates that "people with a comparative advantage for online crime tend to be educated and capable, but they live in societies with poor job prospects and ineffective policing" (Moore et al. 2009).

9.6 Deductions

The application of behavioural science and risk-based decision theories may be useful in providing unique or broad insights into ransomware attackers. Analysis of major ransomware attacks indicates the attackers developed their attacks over an extended period of time for a specific purpose – therefore, the attack is the product of a series of rational and calculated decisions. However, the deployment of the attack may or may not be the result of a rational or calculated decision. At this junction time, the value of this research is unable to be determined, but it does provide a relatively unexplored frontier for understanding cybercrime.

9.7 Conclusion

This chapter began by detailing the profound impact the advent of the Internet has played in society from the creation of new global markets to how we connect and interact with each other. The influence of applied cryptography was fundamental to the evolution of Internet anonymity and encryption, which in turn formed the building blocks for developing and deploying ransomware attacks. The advent of anonymous gift cards and cryptocurrencies influenced how criminals could monetise data and how they could seek payment for their criminal activities. The continued parallel emergence of anonymising technologies has created an environment that has incrementally favoured the attacker, which in turn has triggered an exponential rise in the number and complexity of ransomware attacks.

Cryptocurrencies have played a significant role in the elevation of ransomware becoming a prodigious cyberthreat. They provide a high degree of anonymity for criminals exchanging illicit goods and services online, which aids the further development of ransomware attacks and facilitates RaaS. Arguably, the greatest impact cryptocurrencies have played in the propagation of ransomware attacks is the removal of complicated and profit-depleting money laundering processes. Streamlining the laundering process has exponentially increased the profits attackers could potentially make from undertaking ransomware attacks.[2] The laundering process is further enhanced by the ability of attackers to coerce victims to pay ransoms in cryptocurrencies, which inhibits the tracking of ransomware payments. The advent and adoption of cryptocurrencies signal a new era of anonymous financial transactions beyond the control of financial regulators and law enforcement.

The rapid adoption of connected devices has challenged enterprise security and risk management practices. Shortages in adequately trained cybersecurity and cyber risk professionals have required organisations to take increased risks in internal staff

[2] Note: The potential money laundering advantage is not limited to ransomware attacks or broader cyberattacks. Cryptocurrencies may provide the same streamlining advantages to non-cybercrimes.

appointments. The inherent risk associated with these appointments has been compounded by the complex nature of risk management frameworks that often require specialist cyber expertise and extensive funding for their effective implementation. The convergence of behavioural sciences sub-disciplines with cybersecurity is truly fascinating field that remains relatively unexplored at this point. Presently, there is extremely limited superlative research on the sociology and psychology of ransomware attacks and attackers and broader cyberattacks.

References

L. Ablon, M. Libicki, A. Golay, Projections and predictions for the black market, in *Markets for Cybercrime Tools and Stolen Data Book Subtitle: Hackers' Bazaar*, (RAND Corporation, Santa Monica, 2014a)

L. Ablon, M. Libicki, A. Golay, Characteristics of the black market, in *Markets for Cybercrime Tools and Stolen Data Book Subtitle: Hackers' Bazaar*, (RAND Corporation, Santa Monica, 2014b)

M. Aiken, J. Davidson, P. Amann, Youth pathways into cyber crime. (EUROPOL, 2016). Middlesex, United Kingdom

G. Becker, Crime and punishment: An economic analysis. J. Polit. Econ. **76**, 169–217 (1969)

D. Bernoulli, Exposition of a New Theory on the Measurement of Risk. Econometrica **22**(1), 23–36 (1738)

B. Blaze, The purpose of ransomware. (9 July 2017). Available online: https://bartblaze.blogspot.com/2017/07/the-purpose-of-ransomware.html. Accessed 12 April 2018

R. Briggs, Normative theories of rational choice: Expected utility [Liner notes], in *The Stanford Encyclopedia of Philosophy*, (Stanford University, 2017). Available online: https://plato.stanford.edu/archives/spr2017/entries/rationality-normative-utility/. Accessed 4 Jan 2018

C. Camerer, An experimental test of several generalized utility theories. J. Risk Uncertain. **2**, 61–104 (1989)

J. Coleman, Crime and money: Motivation and opportunity in a monetarized economy. Am. Behav. Sci. **35**, 827–836 (1992)

O. Engdahl, The role of money in economic crime. Br. J. Criminol. **48**, 154–170 (2008)

C. Francescani, Ransomware Hackers Blackmail U.S. Police Departments, *NBC News*. (2016). Available online: https://www.nbcnews.com/news/us-news/ransomware-hackers-blackmail-u-s-police-departments-n561746. Accessed 13 Mar 2018

L. Harris, Rise in child and teen fraud arrests mainly due to increase of Internet-based crimes, *The Daily Telegraph*. 11 Apr 2015 (2015) [Online]. Available online: https://www.dailytelegraph.com.au/news/nsw/rise-in-child-and-teen-fraud-arrests-mainly-due-to-increase-of-internetbased-crimes/news-story/fc620acdb8379e30ab46f17493e40475. Accessed 3 Aug 2019

D. Kahneman, *Thinking, Fast and Slow* (Penguin Books, London, 2011)

D. Kahneman, A. Tversky, Prospect theory: An analysis of decision under risk. Econometrica **47**(2), 263–292 (1979)

A. Kharraz, W. Robertson, D. Balzarotti, L. Bilge, E. Kirda, *Cutting the Gordian Knot: A Look Under the Hood of Ransomware Attacks* (Springer International Publishing, Cham, 2015)

S. Lindenberg, *Rational Choice Theory: Advocacy and Critique*. [eBook] (Sage Publications, Newbury Park, 1992)

G. Mars (ed.), *Cheats at Work: An Anthropology of Workplace Crime*, Ammended edition. (Dartmouthm Aldershot, Hants, England, 1982)

R. McAdams, T. Ulen, *Behavioral Criminal Law and Economics, Working Paper No. 440*, 2008

R. Merton, Social structure and anomie, in *Social Theory and Social Structure*, vol. 190, (The Free Press, New York, 1968)

S. Messner, R. Rosenfeld, *Crime and the American dream*, 2nd edn. (Wadsworth Pub, Belmont, 2001)

T. Moore, R. Clayton, R. Anderson, The economics of online crime. J. Econ. Perspect. **23**(3), 3–20 (2009)

Q. Norton, Anonymous 101: Introduction to the lulz, *WIRED*, (2011). Available online: https://www.wired.com/2011/11/anonymous-101/. Accessed 3 Jan 2017

R. Price, How a hacker got sweet revenge on scammers who tried to take advantage of his parents, *Business Insider*. (2016), 17 June 2018

J. Rabkin, A. Rabkin, Hacking back without cracking up, *Aegis Series Paper*, 1601, (2016),

Trend Micro, 3 Reasons the Ransomware Threat will Continue in 2018. 24 January 2018. Available online: https://blog.trendmicro.com/3-reasons-the-ransomware-threat-will-continue-in-2018/. Accessed 4 June 2018

C. Wood, Insider threat examples: 7 insiders who breached security, *CSO*, (2018). Available online: https://www.csoonline.com/article/3263799/security/insider-threat-examples-7-insiders-who-breached-security.html#slide1. Accessed 10 June 2018

D. Lacey in M. Wordsworth, Cybercrime victims on their own as police fail to follow up cases, helpline head says, *Lateline*. (2017). Available online: http://www.abc.net.au/news/2017-02-27/cybercrime-victims-on-their-own-as-police-fail-to-follow-cases/8306814. Accessed 26 May 2018

Chapter 10
Conclusion

This research has argued that the emergence of new anonymising technologies has enabled ransomware to evolve into a prodigious cyberthreat. Ransomware attacks are a phenomenon that have repeatedly displayed the capacity to rapidly monetise both crown jewels and innocuous data with limited or no monetary value. This has been greatly facilitated by the advent and adoption of cryptocurrencies. Online black markets have facilitated communications between criminals and enabled illegal goods and services to be traded at an unprecedented velocity and level of anonymity. The parallel emergence of multiple encryption-based technologies triggered a propagation of cyber-enabled effects, which has modified how cyberattackers, organised cybercrime syndicates and nation states develop and undertake ransomware attacks.

Encryption is at the crux of cybersecurity. It is used by nation states, businesses and consumers every second of the day to provide security for data at rest, access and during transmission. The application of encryption no longer requires specialist expertise, and the evolution of computer systems has enabled encryption to become easier than ever to apply. The trust we place in strength of encryption and the ease of its deployment is the same for criminals undertaking ransomware attacks. Whilst encryption is fundamental to our cybersecurity, the irony is that it is equally fundamental for providing security and anonymity for cybercriminals. Analysis of ransomware attackers indicates that many have adopted industry best practices to protect themselves and to inhibit conventional law enforcement practices. The use of anonymising technologies such as Tor, encrypted messaging platforms and cryptocurrencies highlights the steps criminals are taking to lower risk and to protect their criminal operations.

The impact cryptocurrencies play in ransomware attacks and broader cybercrimes cannot be underestimated. The advent of cryptocurrencies can be traced to a lack of innovation by central banks, globalisation, government interference and an increasing consumer distrust over the role major financial institutions play in the

M. Ryan, *Ransomware Revolution: The Rise of a Prodigious Cyber Threat*,
Advances in Information Security 85, https://doi.org/10.1007/978-3-030-66583-8_10

global community.[1] The shift towards Internet-based transactions and services has led financial institutions to increase their analytics of consumer data and transactions, and this has continued to erode consumer trust and privacy. These activities led the academic and global Internet community to develop an alternative financial platform which sought transaction transparency whilst increasing the level of consumer privacy.

The launch of Bitcoin signalled the mainstream beginning of cryptocurrencies, which ushered in a new era of anonymous transactions beyond the control of financial regulators and law enforcement. For cybercriminals, cryptocurrencies were the missing link in anonymously trading illicit goods and services online. They provided a method for criminals to securely and instantly transact with other criminals around the globe. This enabled cybercriminals the ability to anonymously trade exploits and transfer the proceeds of crime beyond the purview of law enforcement. It also provided attackers an anonymous way to demand payment for ransomware attacks, removing the need to launder the proceeds. The removal of the requirement to launder the ransomware payments numerous times exponentially increased the profits attackers could make from undertaking ransomware attacks. Cryptocurrencies provided the catalyst for the development and increased profitability of ransomware, sparking a phenomenal rise in the number and complexity of ransomware attacks.

The rapid increase in the volume of Internet-enabled user devices and platforms has directly and inadvertently created an environment that is ripe for undertaking ransomware attacks. In the IoT and OT spaces, the rapid deployment of connected devices across organisations is commonly linked to organisational efforts to increase data gathering points as a way to improve customer operations and to deliver cost savings from fuel, energy and water usage. The B-side of this pursuit for efficiency is the rapid increase in the scale and complexity of enterprise networks. This has drastically increased the number of end points, and the application of Metcalfe's law reveals that large clusters of IoT devices and large enterprise networks are of an exponentially greater value to attackers.

Healthcare providers, critical infrastructure and government agencies remain high-value targets for ransomware attackers. Healthcare providers are considered lucrative targets for attackers because they are reliant on up-to-date information and have limited cyber defences, and their employees have limited cybersecurity training. When ransomware incidents occur in these organisations, they can induce a higher degree of panic because their staff are often already stretched trying to provide vital medical care. Attackers seek to disrupt critical services to exploit this fear in order to coerce ransom payments. The volume of ransomware attacks against these types of organisations and other recent attacks also indicates an increasing transition from generic to targeted ransomware attacks.

[1] Note: A significant driver in distrust occurred in the early 2000s when the US government introduced laws outlawing online gambling. The laws introduction was followed up by agencies pressuring financial institutions to block credit-card transactions to these sites in order to prevent people using the sites.

The analysis also revealed that there is a quandary between professional service providers and academic institutions. As organisations competing for cyber talent continue to increase salaries, many school leavers and professionals entering the cybersecurity industry have opted for alternative educational routes outside of universities. Cybersecurity is an evolving discourse, and it is difficult to predict what long-term impact this shift will have on academia. However, for academic institutions to continue to develop cyber-related course content that is relevant to professional practices, they will become increasingly reliant on students sharing their acquired knowledge and experiences, and this may be problematic in the future. Due to security and contractual restrictions, there are limited data sharing efforts between industry and academia, and as a result, academia has struggled to produce research that engages with business and cybersecurity leaders. For cybersecurity scholars to produce quality cybersecurity research, there is an inherent requirement to have access to raw incident data. This is an area that requires increased data sharing collaborative efforts, and alternative fields such as medical research may provide a suitable model to enable data sharing between enterprises and academic institutions.

Ransomware developers have consistently exhibited the ability to swiftly exploit new vulnerabilities whilst adapting and deploying emerging techniques faster than the forces tasked with preventing them. Advancements in technology and coding resources have lowered the skills entry bar for attackers and enabled advanced encryption techniques to be easily applied to circumvent and inhibit law enforcement operations. This research documented how the use of emerging technologies had made the attribution of major ransomware incidents increasingly difficult. The application and ease of deployment of anonymising technologies complicate the investigation and prosecution processes for ransomware victims and law enforcement agencies. It also adversely impacts deterrence and retaliation strategies that could be utilised by nation states. Even when attribution can be achieved, as it has been previously by the United States and its allies, geopolitics frequently prevented the arrests and prosecution of any criminal suspects. This blocker raises the question, Do states need to develop alternative responses and consequences for states that launch or support ransomware attacks?

The four case studies detailed the continuous evolution in attack methodologies and the growing influence of states in major ransomware attacks. Both advertent and inadvertent states are playing a growing role in the development and deployment of ransomware attacks. The ongoing pursuit of zero-day exploits by states for warfare applications continues to threaten the global cybersecurity environment. Even the management of zero-day exploits has proved challenging for states, with multiple large-scale ransomware attacks utilising re-engineered state-sponsored exploits to infiltrate their victims' networks. It also highlights the limitations and effectiveness of cybersecurity countermeasures against ransomware attacks.

Whilst zero-day exploits represent the upper echelon of threats to the cybersecurity of organisations, more frequently, it has been deficiencies in cybersecurity fundamentals that have been the root cause of many ransomware attacks. Enterprises have been quick to launch new technologies and platforms but have a limited understanding of how to effectively secure those new technologies. The continuous pressure to innovate in order to remain competitive has left many organisations with IT environments that are increasingly complex to manage. The security of legacy infrastructure has often been sacrificed to fund the deployment of new applications and technology, and this has left many large organisations vulnerable to ransomware attacks.

The continued success of ransomware attacks indicates that many governments and corporations lack the required cyber expertise, training and effective expenditure of resources to understand, prepare for and respond to major cyberattacks against their organisations' systems. This research highlighted that the continuous expansion and early adoption of emerging technologies may be beyond the capacity of conventional risk management frameworks and expertise of generic risk managers. The continued periodic application of static risk calculators to determine dynamic threats is problematic. This process is further compounded by a lack of suitably qualified and experienced cyber risk managers. Analysis of the current body of academic research and industry sentiment indicate that these cyber skills deficiencies will most likely continue to increase throughout the next decade.

Looking to future research avenues, the outcomes of this research suggest that ransomware will pose its greatest threat to organisations during periods of escalating tensions between nation states and during periods of cryptocurrency devaluation (including coin halvings). Whilst this research indicates that ransomware has typically been used by criminal syndicates, it also highlighted the use of ransomware by rogue nation states. In periods of escalating tensions, ransomware may provide a unique method to apply non-kinetic force to coerce potential opponents. Ransomware shares many traits associated with asymmetrical warfare, and therefore, it is foreseeable that agencies and special operations may further develop ransomware as a warfare tactic. An additional avenue for future research relates to improving the accuracy of predicting the true financial cost of cyberattacks, and the output of such research would be of significant value to governments and corporations alike.

The transition between criminals stealing and selling users' personal data versus undertaking ransomware attacks or undertaking crypto mining operations is driven purely by profitability. Due to streamlined business processes and agile organisational structures, organised cybercrime syndicates will continue to rapidly adopt and exploit emerging technologies faster than those charged with preventing them. As governments and corporations look to increase their utilisation of cloud-based solutions, attackers will be quietly preparing in the virtual shadows to capitalise on this continued transition. The propensity for security misconfigurations and the concentration of colossal data sets and applications hosted on cloud-based platforms present a potential financial opportunity too great to be ignored for organised cybercrime syndicates and rogue nation states.

Printed in the United States
by Baker & Taylor Publisher Services